"十四五"高等职业教育智能制造系列教材
江苏省高等学校重点教材（编号：2021-2-143）

数控车削编程与加工

（中英双语版）

朱学超 刘 旭 ◎ 主 编
张 良 陈 琪 田 菲 ◎ 副主编

中国铁道出版社有限公司
CHINA RAILWAY PUBLISHING HOUSE CO., LTD.

内 容 简 介

本书以国际化制造业人才培养为目标，通过中英文结合的方式介绍现代最前沿的数控车削编程与加工技术。

本书共分4章，包括数控车床简介、数控车床编程的基本概念、车削固定循环指令、数控加工及程序编制。内容的编排上，理论知识遵循深入浅出，突出数控车削加工技术的先进性和实用性。案例加工以项目形式进行设计，教学的过程体现工作过程。本书选择与企业非常接近的典型零件加工项目，注重学生的岗位职业能力培养，充分体现了"以职业能力为主线，以岗位需求为依据"的教材开发理念。

本书以英文为主，中文作为概括及解释，详略视内容的难易程度而定。文中配有大量插图，以利于提高记忆效果，并减少对中文注释的依赖。同时本书在重要的知识点、技能点上，配有丰富的视频、微课、图片等资源，学习者可以通过扫描其中的二维码获取相关信息。

本书适合作为高等职业院校数控技术、模具设计与制造、机械制造及自动化、机电一体化技术等相关专业双语教学使用，也可作为留学生用教材，并可供相关工程技术人员学习参考。

图书在版编目（CIP）数据

数控车削编程与加工：汉、英/朱学超，刘旭主编．—北京：中国铁道出版社有限公司，2022.10（2024.7重印）
"十四五"高等职业教育智能制造系列教材
ISBN 978-7-113-29157-0

Ⅰ.①数… Ⅱ.①朱… ②刘… Ⅲ.①数控机床-车床-车削-程序设计-高等职业教育-教材-汉、英②数控机床-车床-加工-高等职业教育-教材-汉、英 Ⅳ.①TG519.1

中国版本图书馆CIP数据核字（2022）第087582号

书　　名：	数控车削编程与加工（中英双语版）
作　　者：	朱学超　刘　旭

策　　划：	汪　敏	编辑部电话：（010）63560043
责任编辑：	何红艳　彭立辉	
封面设计：	刘　颖	
责任校对：	安海燕	
责任印制：	樊启鹏	

出版发行：中国铁道出版社有限公司（100054，北京市西城区右安门西街8号）
网　　址：https://www.tdpress.com/51eds/
印　　刷：三河市兴达印务有限公司
版　　次：2022年10月第1版　2024年7月第2次印刷
开　　本：787 mm×1 092 mm　1/16　印张：13.75　字数：352千
书　　号：ISBN 978-7-113-29157-0
定　　价：45.00元

版权所有　侵权必究

凡购买铁道版图书，如有印制质量问题，请与本社教材图书营销部联系调换。电话：（010）63550836
打击盗版举报电话：（010）63549461

前　言

为了全面贯彻国家《关于加快和扩大新时代教育对外开放的意见》，快速推进具有全球视野的高层次国际化人才教育，服务"一带一路"倡议，大力开展留学生的职业教育，本着以就业为导向，以企业岗位操作要领为依据，结合国家职业技能鉴定标准，我们编写了《数控车削编程与加工（中英双语版）》一书。

本书是在编者多年来一直从事双语教学、留学生教学，以及从事数控车削编程与加工教学、科研、生产工作经验的基础上编写的。理论知识遵循深入浅出，遵循知识点的前后衔接，突出数控车削加工技术的先进性和实用性。从系统的角度和学生的认知规律考虑教材内容的编排，使教学内容由浅入深，循序渐进，符合认知规律。本书以企业广泛使用的发那科（FUNC）系统数控车床为例进行讲解，共4章：第1章数控车床简介，第2章数控车床编程的基本概念，第3章车削固定循环指令，第4章数控加工及程序编制。从数控车床简介、编程的基本概念、特殊指令编程方法、典型产品程序编制、产品的加工与尺寸控制，以及附件的加工过程中重要安全防护措施，形成了一个完整系统的产品加工闭环和工程技术创新能力的培养链；加工案例的选取体现知识能力与应用能力相融合，从简单到复杂，从单一到综合，并通过项目教学体现能力发展与职业发展规律相适应、教学过程与工作过程相一致的教学体系和模式，既具有先进性，也具有可读性。同时本书在重要的知识点、技能点上，配有丰富的视频、微课、图片等资源，学习者可以通过扫描其中的二维码获取相关信息。

本书适合作为高等职业院校数控技术、模具设计与制造、机械制造及自动化、机电一体化技术等相关专业双语教学使用，也可作为留学生教材，并可供相关工程技术人员学习参考。本书由苏州市职业大学朱学超、刘旭任主编，张良、陈琪、田菲任副主编。编写分工：朱学超、张良编写了第1章，刘旭、陈琪、田菲编写了第2章，刘旭、张良、朱学超编写了第3章，张良、陈琪、田菲编写了第4章，附录由张良编写。

全书由朱学超负责统筹定稿。

本书的编写,特别感谢苏州市职业大学陈洁、陆春元的支持,校企合作单位苏州圣保利机械制造有限公司为本书提供了典型产品案例,苏州戴尔菲精密机械科技有限公司兑松华高级工程师对本书编写提出宝贵意见,在此一并表示感谢。

由于编者水平有限,书中难免有不足之处,恳请广大读者和同仁提出宝贵意见。

编 者

2022 年 6 月

数控车床操作视频

序号	资源名称	二维码	页码
1	开关机操作		10
2	手动换刀及主轴正反操作		11
3	手动车端面		12
4	外圆刀对刀操作		13
5	切断刀对刀操作		14
6	外螺纹刀对刀操作		15
7	镗刀对刀操作		15
8	内沟槽刀对刀操作		16
9	内螺纹刀对刀操作		16
10	校刀操作		17
11	程序编辑操作		59
12	图形模拟操作		59
13	零件自动加工过程		165

数控车床指令 2D 轨迹清单

序号	资源名称	二维码	页码
1	G00 刀具快速定位指令 2D 轨迹		79
2	G01 直线插补指令 2D 轨迹		77
3	G02 圆弧插补指令 2D 轨迹		80
4	G03 圆弧插补指令 2D 轨迹		81
5	G41、G42 刀尖半径补偿命令（前刀位）2D 轨迹		75
6	G70 精车循环指令 2D 轨迹		111
7	G71 轴向粗车复合循环指令 2D 轨迹		111
8	G72 端面粗加工循环指令 2D 轨迹		112
9	G73 仿形车削复合循环指令 2D 轨迹		112
10	G75 切槽或切断循环指令（切槽）2D 轨迹		130
11	G75 切槽或切断循环指令（切断）2D 轨迹		131
12	G76 螺纹切削复合循环指令 2D 轨迹		146
13	G90 轴向单一循环指令（圆柱切削）2D 轨迹		100

续表

序号	资源名称	二维码	页码
14	G90 轴向单一循环指令(圆锥切削)2D 轨迹		100
15	G92 螺纹切削单一循环指令 2D 轨迹		145
16	G94 端面循环切削指令 2D 轨迹		107
17	M 指令的使用		62
18	刀具补偿原因		71
19	刀具补偿的加入与取消		76
20	数控车对刀 2D 轨迹		12
21	数控车退刀方式		31
22	机床参考点		45
23	机床原点		46
24	编程坐标系		46
25	梯形螺纹加工方法		150
26	外圆车刀的三面二刃一尖		28

续表

序号	资源名称	二维码	页码
27	正交平面参考系		43
28	机床坐标系		43

数控车床指令 3D 视频

序号	资源名称	二维码	页码
1	G01 直线插补指令 3D 视频		80
2	G02 圆弧插补指令 3D 视频		81
3	G03 圆弧插补指令 3D 视频		81
4	G70 精加工循环指令 3D 视频		111
5	G71 外径粗车循环指令 3D 视频		111
6	G72 端面车削固定循环指令 3D 视频		112
7	G73 固定形状粗加工循环指令 3D 视频		112
8	G75 车槽固定循环指令 3D 视频		131
9	G76 复合螺纹车削循环指令 3D 视频		147
10	G90 矩形车削循环指令 3D 视频		100
11	G92 螺纹车削循环指令 3D 视频		145
12	程序的输入方法 3D 视频		59
13	刀具安装方法 3D 视频		26

续表

序号	资源名称	二维码	页码
14	螺纹环规使用方法 3D 视频		146
15	螺纹塞规使用方法 3D 视频		146
16	切槽方法 3D 视频		88
17	双头螺纹车削方法 3D 视频		149

目 录

Chapter 1　Introduction to the CNC Lathe ········ 1
第 1 章　数控车床简介 ········ 1

Learning Objectives ········ 1
学习目标 ········ 1

1.1　Background on the CNC Lathe ········ 2
1.1　数控车床背景介绍 ········ 2
1.2　CNC Lathe Axes of Motion ········ 9
1.2　数控车床坐标轴 ········ 9
1.3　Features of the Machine Control Unit (MCU) Lathe ········ 10
1.3　数控车床的特点及控制单元 ········ 10
1.4　Basic Lathe Operations ········ 21
1.4　数控车床的基本操作 ········ 21
1.5　Tooling for CNC Lathe Operations ········ 25
1.5　数控车床刀具 ········ 25
1.6　Tool Speeds, Feeds, and Depth of Cut for Lathe Operations ········ 28
1.6　数控车床的切削速度、进给量及切削深度 ········ 28
1.7　Feed Directions and Rake Angle for Lathe Operations ········ 31
1.7　数控车床操作过程中的进给方向及前角 ········ 31

Chapter Summary ········ 35
本章总结 ········ 35
Review Exercises ········ 36
回顾练习 ········ 36

Chapter 2　Fundamental Concepts of CNC Lathe Programming ········ 40
第 2 章　数控车床编程的基本概念 ········ 40

Learning Objectives ········ 40
学习目标 ········ 40

2.1　Establishing Locations Via Cartesian Coordinates ……………………………… 42
2.1　通过笛卡儿法则建立坐标系 …………………………………………………… 42
2.2　Types of Tool Positioning Modes ………………………………………………… 43
2.2　刀具定位方式 …………………………………………………………………… 43
2.3　Reference Point, Machine Origin, Program Origin (Fanuc Controllers) …… 45
2.3　机床参考点、机床原点和编程原点(FANUC 控制器) ……………………… 45
2.4　Methodizing of Operations for CNC Lathe …………………………………… 47
2.4　数控车床的操作方法 …………………………………………………………… 47
2.5　Setup Procedures for CNC Lathe ………………………………………………… 51
2.5　数控车床加工前的设置 ………………………………………………………… 51
2.6　Programming Language Format ………………………………………………… 57
2.6　程序格式 ………………………………………………………………………… 57
2.7　Important Preparatory Functions (G Codes) for Lathe ……………………… 61
2.7　数控车床的重要准备功能代码(G 代码) ……………………………………… 61
2.8　Important Miscellaneous Functions (M Codes) for Lathe …………………… 62
2.8　数控车床重要的辅助功能代码(M 代码) ……………………………………… 62
2.9　Setting the Machining Origin …………………………………………………… 63
2.9　建立机床原点 …………………………………………………………………… 63
2.10　Feed Rate (F Code) ……………………………………………………………… 65
2.10　进给功能(F 代码) ……………………………………………………………… 65
2.11　Spindle Speed (S Code) ………………………………………………………… 66
2.11　主轴功能(S 代码) ……………………………………………………………… 66
　　2.11.1　Spindle Speed With Constant Surface Speed Control …………………… 66
　　2.11.1　主轴恒定线速度控制 ………………………………………………………… 66
　　2.11.2　Spindle Speed With Clamp Speed and Constant Surface Speed Controls … 67
　　2.11.2　恒定线速度控制 ……………………………………………………………… 67
2.12　Automatic Tool Changing ……………………………………………………… 69
2.12　自动换刀功能 …………………………………………………………………… 69
2.13　Tool Edge Programming ………………………………………………………… 70
2.13　刀尖编程方式 …………………………………………………………………… 70
2.14　Tool Nose Radius Compensation Programming ……………………………… 71
2.14　刀尖圆弧半径补偿编程方式 …………………………………………………… 71
　　2.14.1　Setting Up Tool Nose Radius Compensation ……………………………… 72
　　2.14.1　设置刀具刀尖圆弧半径补偿 ………………………………………………… 72
　　2.14.2　Some Restrictions With Tool Nose Radius Compensation ……………… 73
　　2.14.2　刀尖圆弧半径补偿的注意事项 ……………………………………………… 73
　　2.14.3　Tool Nose Radius Compensation Commands ……………………………… 74

2.14.3 刀尖圆弧半径补偿指令 ··· 74

2.15 G Code ··· 77
2.15 G 指令 ··· 77
 2.15.1 Linear Interpolation Commands ··· 77
 2.15.1 直线插补指令 ··· 77
 2.15.2 Circular Interpolation Commands ··· 79
 2.15.2 圆弧插补指令 ··· 79
 2.15.3 Grooving Commands ··· 87
 2.15.3 切槽指令 ··· 87

2.16 Return to Reference Point Command ··· 89
2.16 返回机床参考点指令 ··· 89

Chapter Summary ··· 90
本章总结 ··· 90

Review Exercises ··· 92
回顾练习 ··· 92

Chapter 3   Techniques and Fixed Cycles for CNC Lathe Programming ··· 98

第 3 章   车削固定循环指令 ··· 98

Learning Objectives ··· 98
学习目标 ··· 98

3.1 Turning and Boring Cycle: G90 ··· 99
3.1 外圆车削和镗孔循环指令: G90 ··· 99

3.2 Facing Cycle G94 ··· 105
3.2 端面车削循环指令 G94 ··· 105

3.3 Multiple Repetitive Cycles: G70 TO G75 ··· 109
3.3 复合循环指令: G70 ~ G75 ··· 109
 3.3.1 Stock Removal in Turning and Boring Cycle: G71 ··· 109
 3.3.1 粗车及粗镗循环指令: G71 ··· 109
 3.3.2 Finish Turning and Boring Cycle: G70 ··· 111
 3.3.2 精车及精镗循环指令: G70 ··· 111
 3.3.3 Peck Drilling and Face Grooving Cycle: G74 ··· 122
 3.3.3 钻孔和端面切槽循环指令: G74 ··· 122
 3.3.4 Peck Cutoff and Grooving Cycle: G75 ··· 130
 3.3.4 切断和切槽循环指令: G75 ··· 130

		3.4　Thread Cutting on CNC Lathes and Turning Centers ········· 138
		3.4　数控车床及车削中心中的螺纹加工 ··································· 138
			3.4.1　Single-Pass Threading Cycle：G32 ································· 139
			3.4.1　单线螺纹循环指令：G32 ··· 139
			3.4.2　Multiple-Pass Threading Cycle：G92 ······························ 144
			3.4.2　多线螺纹循环指令：G92 ··· 144
			3.4.3　Multiple Repetitive Threading Cycle：G76 ························ 147
			3.4.3　多重复合螺纹循环指令：G76 ······································· 147
	Chapter Summary ·· 152
	本章总结 ··· 152
	Review Exercises ·· 153
	回顾练习 ··· 153

Chapter 4　CNC Machining and Programming ············· 165
第4章　数控加工及程序编制 ······································· 165

	Learning Objectives ··· 165
	学习目标 ··· 165
		4.1　Project 1 Program External Contour ································· 166
		4.1　项目1 外轮廓的数控编程 ··· 166
			4.1.1　Project Import ·· 166
			4.1.1　项目导入 ··· 166
			4.1.2　Project Implementation ··· 166
			4.1.2　项目实施 ··· 166
		4.2　Project 2 Program a Simple Stepped Shaft ························· 169
		4.2　项目2 阶梯轴的数控编程 ··· 169
			4.2.1　Project Import ·· 169
			4.2.1　项目导入 ··· 169
			4.2.2　Project Implementation ··· 169
			4.2.2　项目实施 ··· 169
		4.3　Project 3 Program a Complex Stepped Shaft ······················· 173
		4.3　项目3 复杂阶梯轴数控编程 ·· 173
			4.3.1　Project Import ·· 173
			4.3.1　项目导入 ··· 173
			4.3.2　Project Implementation ··· 174
			4.3.2　项目实施 ··· 174

4.4　Project 4　Program a Shaft with Arcs ……………………………………………… 179
4.4　项目4　带圆弧轴的数控编程 ………………………………………………… 179
　　4.4.1　Project Import ………………………………………………………… 179
　　4.4.1　项目导入 ……………………………………………………………… 179
　　4.4.2　Project Implementation ……………………………………………… 179
　　4.4.2　项目实施 ……………………………………………………………… 179
4.5　Project 5　Program a Threaded Shaft ………………………………………… 184
4.5　项目5　螺纹轴的数控编程 …………………………………………………… 184
　　4.5.1　Project Import ………………………………………………………… 184
　　4.5.1　项目导入 ……………………………………………………………… 184
　　4.5.2　Project Implementation ……………………………………………… 185
　　4.5.2　项目实施 ……………………………………………………………… 185

Appendix A　Important Safety Precautions ……………………………… 191
附录A　重要安全措施 ……………………………………………………………… 191

Appendix B　Summary of G Codes, M Codes and Auxiliary Functions for Turning Operations ……………………………………… 195
附录B　数控车的G代码、M代码和辅助变量汇总 ……………………………… 195

Appendix C　Summary of Speeds and Feeds for Turning …… 202
附录C　车削加工过程中的车削速度和进给量汇总 …………………………… 202

参考文献 ………………………………………………………………………………… 204

Chapter 1　Introduction to the CNC Lathe

第 1 章　数控车床简介

Learning Objectives

At the conclusion of this chapter you will be able to:

(1) Understand the basic elements comprising the CNC lathe.
(2) Identify the axis of motion for CNC lathes.
(3) Describe the most important cutting operations performed on the CNC lathe.
(4) State the different types of tooling used for CNC lathe operations.
(5) Name the important components of CNC lathe tooling.
(6) Explain lathe feeds, speeds, and depth of cut.
(7) Understand the importance using right hand, left hand, neutral tooling, and rake angles for cutting materials.

学习目标

在本章结束时，你将能够：

(1) 了解数控车床的基本组成要素。
(2) 认识数控车床的运动轴。
(3) 描述在数控车床上执行的最重要的切削操作。
(4) 阐述用于数控车床操作的不同类型的刀具。
(5) 说出数控车床工具的重要组成部分。
(6) 了解车床的进给量、切削速度和切削深度。
(7) 理解使用右偏刀、左偏刀、中性刀和刀具倾角的重要性。

This chapter considers the basic elements of modern CNC lathes and turning centers. The lathe axes are discussed and illustrated. The most important lathe operations are examined, together with the type of tooling required for each operation. Speeds, feeds, and depth of cut for lathe work are important both for prolonging tool life and producing a quality cut. Thus, a concise presentation of these parameters and a brief table for various materials are given. The concept of right, left, and neutral tooling is fundamental and a brief discussion is given. The chapter ends with a discussion of rake angles and selecting the proper rake angle for cutting different materials.

本章考虑了现代数控车床和车削中心的基本要素。对车床运动轴进行了论述和说明。对最重要的车床操作以及每种操作所需的刀具类型进行研究。车削加工速度、进给量和切削深度对延长刀具寿命和加工质量有重要意义。因此，本章简明阐述了这些参数，并给出了各种材料适用的加工参数，给出了右偏、左偏、中性刀具的基本概念并进行了简单的讨论。本章最后论述了前角和选择合适的前角来切削不同的材料。

1.1　Background on the CNC Lathe

1.1　数控车床背景介绍

1. What Is the CNC Lathe

1. 什么是数控车床

The CNC lathe is a machine tool designed to remove material from stock that is clamped and rotated around the spindle axis. Most metal cutting is done with a sharp single-point cutting tool. Drilling, reaming, tapping, turning, and boring are operations performed on a CNC lathe. Modern CNC lathes use a chuck attached to the spindle. The chuck grips and rotates the work. A turret mechanism is often used to hold and index cutting tools are called by the word address part program. Some machines employ two turrets: front and rear. Tools may also be mounted on the shield. A front turret is built to move tools from below the spindle centerline up to the work. A rear turret, on the other hand, moves tools from above the spindle centerline down to the work. Machines equipped with both front and rear turrets can execute cutting operations simultaneously from above and below the work. The modern CNC Lathe refer to Figure 1-1.

数控车床是一种机床，旨在从夹紧并绕主轴轴线旋转的原料中切出物料。大多数金属切削是用锋利的切削工具完成的。钻孔、铰孔、攻丝、车削和镗削都是在数控车床上进行的操作。现代数控车床采用一个与主轴相连的卡盘。卡盘夹紧并旋转工件。一个转台机构经常通过字地址程序来保存和更换刀具。一些车床会在前面和后方使用两个刀架。刀具也可以安装在垫片上。前置刀架是为将刀具从主轴中心线下方移动到工件上。而后置刀架将刀具从主轴中心线上方向下移动到工件上。配备有前后刀架的机床可同时从上方和下方执行切削操作，现代数控车床如图1-1所示。

Figure 1-1　The Modern CNC Lathe
图 1-1　现代数控车床

Chapter 1 Introduction to the CNC Lathe
第1章 数控车床简介

A two-spindle-opposed-type CNC lathe is composed of two spindles facing each other and two turrets. It is capable of machining the front and the back of the part by moving it from one spindle to the other.

双轴对立型数控车床是由两个相互对置的主轴和两个刀架组成,并且能够通过将工件从一个主轴移动到另一个主轴来加工零件的正面和背面。

The CNC vertical turret lathe is similar to a horizontal machine, but the spindle is vertical and perpendicular to the chuck. It is designed to machine work that is large and/or heavy. The automatic screw machine is an automated turret lathe especially designed to produce high-volume, low-cost parts. A metal rod is fed into the machine and turned to meet the manufacturer's specifications. A "live tooling" mechanism attached to the turret has its own motor and spindle for holding and rotating milling tools. Lathes equipped with live tooling can tap side holes, mill hexes, flats, or similar patterns on the part circumference while the lathe spindle remains locked, which adds flexibility by reducing setup time for secondary operations.

数控立式转塔车床类似于卧式车床,但机床主轴是竖立的并且垂直于卡盘。这是被设计用于加工大的或重的工件。自动螺丝车床是一种自动转塔车床,专为生产大体积和低成本的零件而设计的。一根金属棒料安装在车床上并被转动加工成符合制造商要求的零件。安装在转塔上的"活工装"机构自配电动机和主轴,用于夹持和旋转刀具。装有带电工具的车床可以在车床主轴保持锁定的同时在零件圆周上攻丝侧孔、铣六角形、平面或类似的图案,通过减少二次操作的安装时间而增加了灵活性。

2. General Safety Rules for Leather Operations
2. 机床操作安全规则

(1) Prior to operation, confirm that there are no other personnel in the immediate vicinity of the machine.

(1)在操作之前,确认机库附近没有其他人员。

(2) Be sure the machine has completely stopped before entering work area.

(2)在进入工作区域之前确保机床已经完全停止。

(3) As shown in Figure 1-2, wear ANSI-approved safety goggles. Secure all loose clothing. Put long hair up.

(3)如图 1-2 所示,佩戴 ANSI 认可的安全护目镜,不要穿宽松的衣服,把长发盘起来。

(4) Observe the following practices when cleaning the machine:

Figure 1-2 Wear ANSI-approved Safety Goggles
图 1-2 佩戴 ANSI 认可的安全护目镜

- Clean the spindle bore and collets.
- Use a wire brush to remove foreign particles from the threads.
- Do not use an air hose.

(4)在打扫机床时观察下列操作:
- 清洗主轴孔和夹头。
- 使用钢丝刷从螺纹中去除异物。
- 不要使用空气软管。

(5) Check all seating areas on the tool holder to be sure each tool rests solidly.

(5)检查刀架上的所有工位区域,确保每个刀具安置牢固。

(6) Revolve the spindle by hand before starting any operation. Check to see that the chuck jaws work.

(6)在开始任何操作之前,用手旋转主轴。检查卡盘爪是否工作。

(7) Avoid excessive overhang of the cutting tool or boring bar.

(7)避免刀具或镗杆太过悬空。

(8) Keep area around machine clean, dry, and free of obstructions, as shown in Figure 1-3.

(8)保持机床周围干净、干燥、无障碍物,如图1-3所示。

Figure 1-3　Keep Area Around Machine Clean, Dry, and Free of Obstructions
图1-3　保持机床周围干净、干燥、无障碍物

3. Components of CNC Lathes
3. 数控车床组件

CNC lathes have the following primary components: bed, headstock, chunk, turret, carriage, tailstock, slant bed and machine Control Unit (MCU).

数控车床有以下主要部件：

(1) Bed

(1) 床身

The bed aligns and rigidly supports the turret, tailstock, ways, and other key components of the lathe. The bed is designed to transfer vibrations away from the cutting area. Most CNC lathes have slanted beds to allow chips and coolant to fall away easily.

床身对齐并牢固支撑刀塔、尾架、导轨，以及车床的其他关键部件。床身的设计是为了使振动远离切削区域。大多数数控车床都是倾斜的床身，以便于切屑和冷却液的脱落。

(2) Headstock

(2) 主轴箱

The spindle and gear transmission system for rotating the cutting material are contained in the headstock. A variable speed motor drives the spindle.

在主轴箱中包含主轴和齿轮传动系统，此系统用于旋转切削材料。可调速电动机用于驱动主轴。

(3) Chuck

(3) 卡盘

The chuck is connected to the spindle, where clamps and rotates the work. A manually operated chuck is tightened by a chuck wrench, and a power chuck is controlled by a foot switch. The three jaw "self-centering" chuck moves all of its jaws simultaneously to clamp or unclamp the work, and it is used for work with a round or hexagonal cross-section, as shown in Figure 1-4.

卡盘与主轴相连，用于夹紧和旋转工件。手动操作的卡盘通过卡盘扳手拧紧，动力卡盘由脚踏开关控制。三爪"自定心"卡盘能够同时移动其所有卡爪以夹紧或松开工件，并用于圆形或六边形横截面的切削，如图1-4所示。

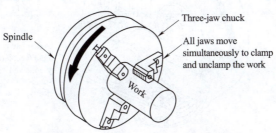

Figure 1-4　The Three Jaw "Self-centering" Chuck

图1-4　三爪"自定心"卡盘

A four "independent jaw" chuck can clamp on the work by moving each jaw independent of the others. This chuck exerts a stronger hold on the work and it has the ability to center non round shapes (squares, rectangles) exactly, as shown in Figure 1-5.

四个"独立夹爪"夹头可以通过单独移动每个夹爪来夹紧工件。这种夹头可以更好地固定工件，并且能够准确地将非圆形（正方形、矩形）工件位于居中位置，如图1-5所示。

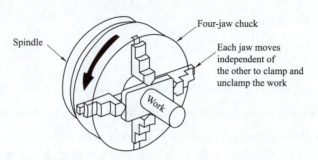

Figure 1-5 A Four "Independent Jaw" Chuck

图 1-5 四个"独立夹爪"夹头卡盘

A lathe faceplate is used for work of irregular shape that cannot be held in a chuck or mounted between centers. It is mounted on the headstock of the lathe instead of the chuck. A fixture or any other means of holding the part to the face plate is used. As show in Figure 1-6, the work is attached to an angle plate by bolts, and the angle plate is fastened to the faceplate by bolts and T-nuts. Counterweights must be used to offset the throw when a heavy piece of work, such as an angle plate, is mounted off center in order to reduce vibration and chatter, and to guarantee a round shape.

花盘式夹具用于不能固定在卡盘中或安装在中心之间的不规则形状的工件。它安装在车床的主轴箱上而不是卡盘上。夹具或其他固定装置将工件固定在花盘上。如图 1-6 所示,工件通过螺栓连接到角板,并且角板通过螺栓和 T 形螺母固定到面板上。当重型工装(如角板)被安装在偏离中心时,必须使用配重来保持平衡,以减少振动和颤抖,并保证其形状为圆形。

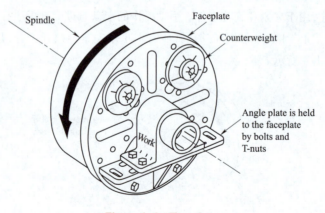

Figure 1-6 Faceplate

图 1-6 花盘

The two main types of jaws used with the chuck are hard jaws and soft jaws, as shown in Figure 1-7.

夹头使用的两种主要类型:硬爪和软爪,如图 1-7 所示。

Hard jaws are used to apply maximum gripping to unfinished surfaces where jaw marks are permissible.

硬爪用于对允许有装夹痕迹的未加工表面施加最大夹持力。

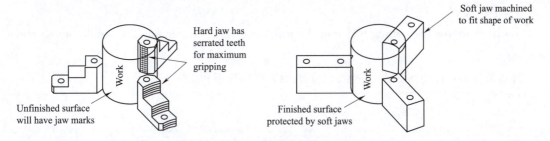

Figure 1-7　Hard Jaws and Soft Jaws
图 1-7　硬爪和软爪

Soft jaws are machined from mild steel to exactly fit the outside shape of the work. They are used for operations where runout must be controlled and to protect finished surfaces when clamping on them.

软爪由低碳钢加工而成，与工件的外部形状完全吻合。它们用于必须控制径向跳动的加工，并在夹紧它们时保护成品表面。

A collet chuck is used when high precision and speed are needed for work of a small cross section. Cross sections normally do not exceed a diameter of 1 in. (1 in = 2.54 cm)

弹簧夹头用于高精度、高速度加工要求的小截面工件，截面直径通常不超过 1 in。（1 in = 2.54 cm）

A feed system (Figure1-8) usually feeds the bar stock. Collet chuck shapes include round, hexagonal, square, and custom shapes.

进料系统（图 1-8）通常用于棒料的进给。夹头形状有圆形、六边形、方形和自定义形状。

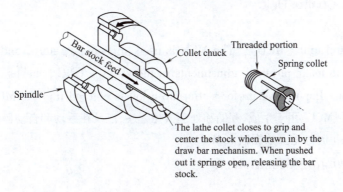

Figure 1-8　A feed System
图 1-8　进料系统

（4）Turret

（4）刀塔

Turret holds the cutting tool and replaces an old tool with a new tool (indexes) during a tool change.

刀塔在换刀期间，夹持刀具并用新刀具替换旧刀具。

(5) Carriage

（5）溜板箱

Carriage moves the cutting tool into the revolving work, it contains the saddle, cross slide, and apron.

溜板箱将刀具移动到旋转工件中，它包含床鞍、横向拖板和床帷。

(6) Tailstock

（6）尾座

The tailstock supports the end of work that is long or has low stiffness and tends to flex away from the cutting tool during machining. Manual and programmable tailstocks are available on CNC lathes. The tailstock moves along its own way with a center to the work. Most CNC lathes use tailstocks with a "live center." A live center has the 60° taper end that fits into the work free to rotate on ball bearings. Friction is greatly reduced and lubrication is not required.

尾座用于支撑较长的工件加工或刚度较低的材料加工，并且在加工期间有效防止工件变形。数控车床上提供手动和可编程的尾座。尾座按自己的方式移动，中心位于工件上。大多数数控车床使用带有"活动中心"的尾座。活动中心具有60°锥度的末端，可以在滚珠轴承上自由旋转。摩擦力大大降低，不需要润滑。

(7) Slant Bed

（7）斜床身

Slant bed supports all the components listed and provides a path for chips and lubricant as they fall.

斜床身支撑以上列出的所有组件，并提供切削和润滑剂滑落的路径。

(8) Machine Control Unit

（8）机床控制单元

Machine control unit (MCU)—The MCU is used to generate, store, and process CNC programs, as well as to make physical adjustments in the way the CNC behaves during manual and automatic operation. Important data for setting up a job is entered into the MCU.

机床控制单元(MCU)用于生成、存储和处理数控程序，以及在手动和自动操作期间对数控的操作进行调整。将加工时的重要数据输入MCU。

4. Tool-Changing Mechanism

4. 换刀机制

A turret on a CNC lathe is used to quickly change tools when required. Each tool position on the turret is numbered to identify the tool it holds. Tools can be mounted on the face of the turret as well as on its sides. Upon receiving a tool change command from the MCU, the turret moves to a safe tool change location and indexes the old tool out and the new tool in, then it proceeds to the proper coordinates programmed for cutting the part with the new tool. For turning centers with programmable tailstocks, the tailstock may have to be moved back before a tool change is executed. The motion of the turret during atypical tool change operation is illustrated in Figure 1-9.

数控车床上的刀塔用于在需要时快速更换刀具。转盘上的每个刀具位置都有编号,用于识别它所固定的刀具。刀具可安装在刀塔平面或侧面上,从 MCU 接收到换刀命令后,刀塔移动到安全的换刀位置,根据指令用新刀具替换旧刀具,然后进入设定好的坐标,以便用新刀具切削零件。对于具有可编程尾座的车削中心,在执行换刀之前可能必须将尾座向后移动。典型换刀操作期间刀塔的运动如图 1-9 所示。

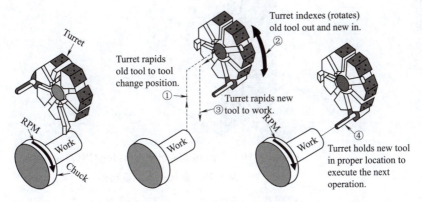

Figure 1-9　Turret Motion During a Tool Change
图 1-9　换刀时的刀塔运动

1.2　CNC Lathe Axes of Motion
1.2　数控车床坐标轴

In this text we will only consider programming basic two – axis machine motions when dealing with CNC lathes. The two-axis are Z and X. The Z axis is in the direction of the spindle. Positive is motion away from the spindle and work and negative $-Z$ is motion toward the spindle and into the work. The X axis controls the cross – slide movements. Positive $+X$ is away from the spindle centerline and negative $-X$ is motion toward the spindle centerline and into the work. Refer to Figure 1-10. In some machines with programmable tailstocks, the W axis is used to designate the movement of the tailstock which accounts for a third axis on these machines. More complex turning centers may have a fourth axis.

接下来我们将只考虑数控车床编程时的两轴机床运动。这两个轴是 Z 轴和 X 轴,Z 轴与主轴轴向重合,Z 轴的正方向是远离主轴和工件的方向,Z 轴的负方向是朝向主轴并运动进入工件的方向。X 轴控制横向移动,X 轴的正方向远离主轴中心线,X 轴的负方向是朝向主轴中心线并进入工件的方向,如图 1-10 所示。在一些具有可编程尾座的机床中,W 轴用于指定尾座的运动,是机床上的第三轴。更复杂的车削中心可以具有第四轴。

视频
开关机操作

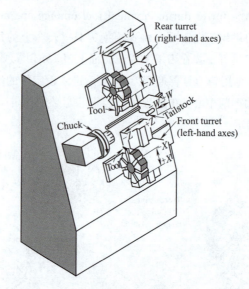

Figure 1-10　Typical Machine Axes for CNC Lathes
图 1-10　数控车床的运动轴

1.3　Features of the Machine Control Unit（MCU）Lathe
1.3　数控车床的特点及控制单元

Specific details and features of machine control units vary from manufacturer to manufacturer. This section provides a genetic presentation of what is found on most machine control units for lathes. Remember to consult the machine tool builder's manual for the detailed information relating to a particular machine control unit used.

机床控制单元的具体细节和特征因制造商而异,本节描述大多数车床控制单元的特点。查阅机床制造商手册,可以了解所使用的特定机床的控制单元的相关详细信息。

In larger companies, it is the CNC setup person, not the programmer, who sets up tooling, loads the job, and runs the first piece on the CNC machine. In smaller operations, the programmer can be expected to get involved in some or all of these tasks. To be more versatile, the programmer needs to acquire a basic knowledge of the features of the machine control unit.

在较大的公司中,不是程序员,而是数控操作工安装刀具和工件,并在数控机床上运行第一件工件。一些简单操作时,程序员可以部分参与或全部参与。为了掌握更全面的技术,程序员需要掌握机床控制单元的基本知识。

The machine control unit is divided into two types of operation panels：the control panel and the machine panel.

机床控制单元有两种类型的操作面板:控制面板和机床面板。

The machine panel is designed and built by the machine tool builder. It contains buttons and switches for controlling the physical behavior of the CNC machine tool. Power buttons turn the CNC lathe on and off. An emergency stop button stops all machine motions. The operator moves a machine axis manually by turning a hand wheel. Control panel can override programmed spindle speeds and feeds, etc.

机床面板由机床制造商设计和制造。它包含用于控制数控机床物理操作状态的按钮和开关。电源按钮可开启和关闭数控车床。紧急停止按钮可停止所有机床运动。操作人员通过转动手轮可手动移动机床运动轴。控制面板可以重新编程设置主轴转速和进给量等。

The operator powers (Figure 1-11) the control panel on or off by depressing these buttons.

操作人员通过按下控制面板开/关电源键(图 1-11)来打开或关闭控制面板。

视频

手动换刀及主轴正反操作

Figure 1-11   Control Panel on/off Power Key

图 1-11   控制面板开/关电源键

Depressing a soft or function key (Figure 1-12) causes a screen menu to appear on the CRT display. The screen menus are used for executing specific functions such as checking a program, displaying the axis position of the machine or entering tool offsets.

按下功能键(图 1-12)之后,屏幕菜单出现在 CRT 显示屏上。屏幕菜单用于执行特定功能,例如检查程序,显示机器的轴位置或输入刀具补偿。

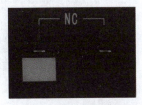

Figure 1-12   Soft or Function Key

图 1-12   功能键

Depressing the axis position key (Figure 1-13) causes the CRT to display the current position of the axis of the CNC machine.

按下位置键(图 1-13)可使 CRT 显示数控机床运动轴的当前位置。

Figure 1-13   The Axis Position Key

图 1-13   位置键

Depressing program key (Figure 1-14) allows the operator to create new word address part programs or see and edit existing programs at the CRT display.

按下程序键(图1-14)允许操作员创建新的字地址零件程序或在CRT显示屏上查看和编辑现有程序。

手动车端面

Figure 1-14　The Program Key

图 1-14　程序键

The tool offset key (Figure 1-15) is used in combination with the cursor position keys to enter the tool offset page and place the cursor at a particular offset value that is to be edited.

刀具偏移键(图1-15)与光标位置键组合使用,打开刀具补偿页面并将光标置于要编辑的补偿值位置。

Figure 1-15　The Tool Offset Key

图 1-15　刀具偏移键

Adress and number keys (Figure 1-16) enable the operator to enter alphabetic and numeric characters. It should be noted that some control panels only contain those alpha keys needed to create word address part programs. An example would be the keys N, G, X.

地址和数字键(图1-16)允许操作员输入字母和数字字符。应注意,一些控制面板仅包含创建字地址零件程序所需的那些字母键,例如键N、G、X。

数控车对刀
2D轨迹

Figure 1-16　Adress and Number Keys

图 1-16　地址和数字键

The operator presses the input key (Figure 1-17) to enter data, like tool offsets, into the control's memory.

操作员按下输入键(图1-17)可将数据(如刀具补偿值)输入控制器的内存。

Figure 1-17　The Input Key

图 1-17　输入键

Pressing one of cursor position keys (Figure 1-18) positions the blinking cursor (left, right, up or down) on the CRT display. Data will be entered at the current cursor position.

按下光标位置键(图 1-18)其中一个按钮,位于 CRT 显示器上闪烁的光标(左、右、上或下)就会显示出来。数据将在当前光标位置输入。

视频

外圆刀对刀操作

Figure 1-18　Cursor Position Keys

图 1-18　光标位置键

The operator presses the program insert key (Figure 1-19) to store a block (one line) of word address code in the control's memory.

操作员按下程序插入键(图 1-19)将一段或一行字地址代码存储在控制器的存储器中。

Figure 1-19　The Program Insert Key

图 1-19　程序插入建

The operator presses the reset key (Figure 1-20) to return the cursor to the start of the program when editing. The key is also used to stop the execution of a part program that has a problem. When this happens, all the program's commands in the look-ahead buffer will be cleared. It should be noted that if the program is executed immediately after reset is pressed, problems may arise as it will run with the commands in the look-ahead buffer skipped. Pressing this key when the CNC is running in alarm state will cancel the alarm.

操作员按下重置键(图 1-20)可在编辑时将光标返回到程序的开头。该键还用于停止执行有问题的程序。发生这种情况时,将清除先行缓冲区中所有程序的命令。应该注意的是,如果在按下复位后立即执行程序,则可能会出现问题,因为它会在跳过前面缓冲区中命令的情况下运行。数控机床在报警状态下运行时按此键将取消报警。

Figure 1-20　The Reset Key

图 1-20　重置键

The operator presses the edit key (Figure 1-21) to set the switch to EDIT mode enables the operator to insert, save, delete, and edit word address part programs at the CRT display using the control panel keypad. The operator can also move to a particular line in a program and begin executing it from that point.

将开关设置为 EDIT 模式（图 1-21），操作人员可以使用控制面板键盘在 CRT 显示屏上插入、保存、删除和编辑字地址零件程序。操作人员也可以移动到程序中的特定行并从该点开始执行。

Figure 1-21　The Edit Key

图 1-21　编辑键

When the switch is set to Manual Data Input or MDI mode by pressing the MDI key (Figure 1-22), setup data like tool length offsets, spindle speed for edge finding and fixture offsets can be entered in. Pressing the START CYCLE button will direct the control to enter the data into memory.

按下 MDI 键将数控车床设置为手动数据输入或 MDI 模式（图 1-22），此时可以输入设置数据，如刀具长度补偿值、寻边和夹具补偿时的主轴转速。按下 START CYCLE 按钮，控制器将数据存入存储器。

视频

切断刀对刀操作

Figure 1-22　The MDI Key

图 1-22　手动输入模式键

JOG mode key (Figure 1-23) puts the CNC machine tool into a manual mode of operation. In this mode the operator can use several devices on the machine panel to manually control its movement. For example, motion along each of the machine axis can be controlled by pressing the + or – JOG buttons. Spindle speed can also be regulated.

寸动模式键使数控机床进入手动操作模式（图 1-23）。在此模式下，操作人员可以使用机床面板上的多个按钮手动控制机床的运动。例如，可以通过按"＋"或"－"寸动按钮控制机床轴的运动。主轴转速也可调节。

Figure 1-23　The Jog Mode Key

图 1-23　寸动模式键

The handle mode key (Figure 1-24) enables the operator to manually jog each of the machine axes by turning the hand wheel on the machine panel.

手轮模式键（图 1-24）可使操作人员通过手轮来控制每个机床轴的运动。

Figure 1-24　The Handle Mode Key

图 1-24　手轮模式键

The operator presses the cycle start key (Figure 1-25) to direct the CNC machine tool to execute a part program selected from the control's memory. Execution will occur in automatic cycle mode.

操作人员按下循环启动键(图1-25)控制数控车床执行从控制器存储器中选择的程序,执行程序将在自动循环模式下进行。

视频

外螺纹刀
对刀操作

Figure 1-25　The Cycle Start Key

图 1-25　循环启动键

The feed hold key (Figure 1-26) is used to cause a momentary halt in the movement of the CNC machine axis. This is done, for example to remove chips, correct coolant flow or prevent a problem from occuring during the execution of the program. If the program needs to be edited, the operator presses the RESET key to cancel its execution. Pressing the cycle start key takes the CNC out of tied hold.

进给保持键(图1-26)用于暂时停止数控机床轴的运动。这样做是为了去除切屑,校正冷却液喷射方向或防止在程序执行期间出现问题。如果需要编辑程序,操作员按 RESET 键取消执行。按循环启动键可使数控退出进给保持。

视频

镗刀对刀
操作

Figure 1-26　The Feed Hold Key

图 1-26　进给保持键

The programmed feed rate, F, can be changed by using the feed override switch (Figure 1-27) when commands to cut such as G1, G2, G3, are executed. Normal increments are 10%. The feed rate can be adjusted from 0, or no feed, to 200%, or double the feed rate. Note, this switch has no effect on the rapid movements of the CNC machine tool.

当执行要切削指令(例如 G1、G2、G3)时,可以使用进给倍率调整开关(图1-27)更改编程时定义的进给倍率 F，正常增量为 10%。进给速率可以从 0 或无进给调节至 200% 或进给速率的两倍。请注意，此开关对数控机床的快速运动没有影响。

●视频
内沟槽刀对刀操作

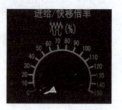

Figure 1-27　The Feed Override Switch

图 1-27　进给倍率调整开关

The programmed spindle speed, S, can be changed by using the spindle override switch (Figure 1-28) when commands to turn the spindle on clockwise (M3) or counterclockwise (M4) are executed. Normal increments are 10%. The switch starts at 50%, for safety, and does not allow for a zero spindle speed override. This could cause a crash if the corresponding feed override were set high.

当执行顺时针（M3）或逆时针（M4）打开主轴的命令时，可以使用主轴倍率调整开关（图 1-28）更改编程时定义的主轴转速 S，正常增量为 10%。为了安全，开关从 50% 开始，不允许主轴转速倍率为零。如果相应的进给量设置为较高时，则可能导致崩溃。

●视频
内螺纹刀对刀操作

Figure 1-28　The Spindle Override Switch

图 1-28　主轴倍率调整开关

Control over the rapid rate of motion toward or away from the work is useful when testing a program containing G0, G27, G28, G29, or G30 commands. The rapid rate can be adjusted to 0, 25%, 50%, 75% or 100% of its normal value by using the rapid override switch (Figure 1-29).

在使用包含 G0、G27、G28、G29 或 G30 命令的程序时，控制朝向或远离工件的快速移动是有效的。使用快速移动倍率调整开关（图 1-29）调整速率为其正常值的 0%、25%、50%、75% 或 100%。

Figure 1-29　The Rapid Override Switch

图 1-29　快速移动倍率调整开关

The operator uses the axis select switch (Figure 1-30) to select the machine axis (X or Z) that is to be jogged by using the JOG buttons or the hand wheel.

操作员使用运动轴选择开关(图 1-30)选择寸动或手轮控制的机床轴(X 轴或 Z 轴)。

The rate switch (Figure 1-31) controls the distance (see Table 1-1) the machine moves along an axis with a single push of the jog button or a single click of the hand wheel.

倍率开关(图 1-31)控制按下寸动键或旋转手轮一格机床沿轴移动的距离,见表 1-1。

Figure 1-30　The Axis Select Switch

图 1-30　运动轴选择开关

Figure 1-31　The Rate Switch

图 1-31　倍率开关

视频

校刀操作

Table 1-1　Moving Rate

Rate	Movement Along a Machine Axis Caused by Asingle Button Push or Wheel Click	
	Operatiing in Metric Mode	Operatiing in Inch Mode
×1	0.002 54 mm	0.0001 in
×10	0.025 4 mm	0.001 in
×100	0.254 mm	0.01 in

表 1-1　移动倍率

速率	按下一次寸动键或旋转手轮一格,车床沿某一运动轴的运动量	
	米制模式	英制模式
×1	0.002 54 mm	0.0001 in
×10	0.025 4 mm	0.001 in
×100	0.254 mm	0.01 in

The mode switch must be set to the handle mode before the hand wheel (Figure 1-32) can be used. The axis select switch selects the axis to jog. Wheel rotations clockwise cause electrical pulses that produce ＋ axis CNC movement. Counterclockwise rotations result in-axis CNC machine movement. The hand wheel is very useful for moving the machine axis during setup. The hand wheel is often mounted on a pendent making it more convenient for the setup person to move around when working.

在使用手轮(图 1-32)之前,必须将模式开关设置为手轮模式。在选择运动轴开关中选择要进行点动的轴。顺时针方向旋转会产生电脉冲,产生运动轴向正方向移动。逆时针旋转会产生运动轴向负方向移动。手轮对于在加工前准备期间移动机床运动轴非常有用。手轮通常悬挂在机床上,使得操作人员在工作时更方便移动。

Figure 1-32　The Hand Wheel
图 1-32　手轮

When the dry run key (Figure 1-33) is set to ON, the feed rate, F, as specified in the running part program is ignored by the control. Instead, the machine axes are moved at higher feed rates. Also, the rapid feed rate can be controlled by the rapid override switch. This is usually done when testing a new part program with the part not mounted.

当空运行键(图1-33)设置为打开状态,控制器忽略运行零件程序中指定的进给速率F,而机床轴以更高的进给速率移动。此外,快速移动速度可由快速倍率调整开关控制。这通常在测试未安装工件的新零件程序时使用。

Figure 1-33　The Dry Run Key
图 1-33　空运行键

Setting the single block key (Figure 1-34) to ON directs the control to execute a single block of the part program and then stop the CNC machine by canceling the AUTO cycle. To execute the next block in the program the operator must press the CYCLE START button, etc. This is usually done to carefully check the machine motions when running a new part program.

将单段键(图1-34)设置为打开状态,可指示控制器执行零件程序的单个程序段,然后通过取消循环加工来停止数控机床。要执行程序中的下一个程序段,操作员必须按下循环启动(CYCLE START)按钮。这通常用于在运行新零件程序时,检查机床的运动状态。

Figure 1-34　The Single Block Key
图 1-34　单段键

The operator sets the machine lock key (Figure 1-35) to ON to step movement along any machine axis when a part program is executed by the controller. When in effect, the CRT will display the program movement along the axes, the spindle will run and other machine functions will operate normally. Machine lock is used to check a new part program.

当控制器执行零件程序时,操作人员将机床锁止键(图1-35)设置为打开状态,控制机床的运动轴进行移动。实际上,CRT 将显示该程序中轴的运动,主轴将运转,其他机器功能将正常运行。机器锁止用于检查新的零件程序。

Figure 1-35　The Machine Lock Key
图 1-35　机床锁止键

Any block in a part program that has a slash (/) as the first character will not be executed when optional block skip key (Figure 1-36) is set to ON.

当跳步键(图 1-36)设置为打开时,将不会执行零件程序中以斜杠(/)作为第一个字符的任何块。

Figure 1-36　Optional Block Skip Key
图 1-36　跳步键

When optional block stop key (Figure 1-37) is set to ON, the controller will slop the execution of a part program and take the control out of AUTO cycle when a M01 block is executed in the part program. The operator presses the CYCLE START button to start the machine again in AUTO mode. If this key is OFF, the control will ignore any programmed M01 codes.

当选择停止键(图 1-37)设置为打开时,该控制器将使零件程序的执行变慢,并在零件程序中执行 M01 程序段时将控制器退出循环。操作人员按下循环启动按钮,在 AUTO 模式下再次启动机床。如果此开关为关闭,则控制器将忽略任何已编程的 M01 代码。

Figure 1-37　Optional Block Stop Key
图 1-37　选择停止键

The tailstock forward/retract keys (Figure 1-38) are active when the mode switch is set to JOG mode.

(1) Pressing TAILSTOCK FOR causes the tailstock to be manually moved forward by jogging it toward the spindle.

(2) Pressing TAILSTOCK RET causes the tailstock to be manually retracted by jogging it away from the spindle.

当模式开关设置为寸动模式时,尾座向前/向后键(图 1-38)处于激活状态。
(1) 按下尾座前移会使尾座按照点动方式向主轴运动。
(2) 按下尾座后退会使尾座按照点动方式远离主轴运动。

Figure 1-38　The Tailstock Forward/Retract Keys

图 1-38　台尾向前/台尾向后

The turret index direction key/turret index key (Figure 1-39) is active when the mode switch is set to JOG mode

(1) The operator first selects the direction in which the turret is to index by pressing one of the direction keys:◁▷

(2) The turret is then jogged in the direction selected by pressing the turret index direction key.

当模式开关设置为寸动模式时,手动选刀键(图1-39)处于激活状态。

(1) 操作人员首先通过按下方向键◁▷选择旋转刀塔的索引方向。

(2) 然后按下换刀键,使刀塔按所选方向移动。

Figure 1-39　The Turret Index Direction Key/Turret Index Key

图 1-39　手动选刀键

The EMERGENCY STOP (Figure 1-40) button should only be pressed when it is necessary to immediately stop between the tool and the work.

(1) Power to the CNC machine will be turned off. Control power will remain.

(2) The control unit will automatically be RESET

(3) After fixing the problem, RESET the CNC by turning the EMERGENCY STOP button and releasing it.

(4) Press the CYCLE START button to start the program again from the beginning.

当需要立刻停止时,才应按下紧急停止按钮(图1-40)。

(1) 将关闭数控机床的电源,控制电源将保持不变。

(2) 控制单元将自动复位。

(3) 解决问题后,通过转动急停按钮并释放它来重置数控机床。

(4) 按循环启动按钮从头开始重新启动程序。

Figure 1-40　Emergency Stop Button

图 1-40　紧急停止按钮

1.4　Basic Lathe Operations

1.4　数控车床的基本操作

　　This text will consider programming the most basic lathe operations: facing, turning, grooving, parting, drilling, boring, and threading. Some of these cuts are performed on both the outside surface of the part (OD operations) as well as the inside surface (ID operations). These are illustrated in Figure 1-41 ~ Figure 1-47.

　　这里将介绍编程最基本的车床操作：车端面、车外圆、车槽、切断、钻孔、镗孔和车螺纹。这些切削动作可以在工件的外表面以及内表面上进行，如图 1-41 ~ 图 1-47 所示。

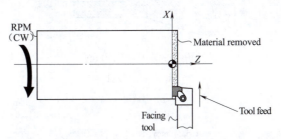

Figure 1-41　Facing

图 1-41　车端面

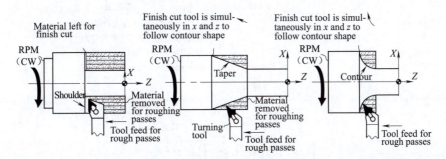

Figure 1-42　Turning

图 1-42　车外圆

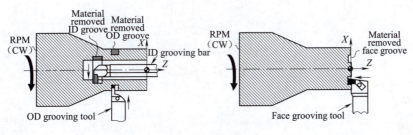

Figure 1-43　Grooving

图 1-43　车槽

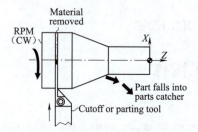

Figure 1-44 Parting

图 1-44 切断

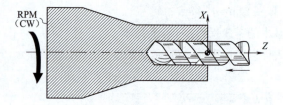

Figure 1-45 Drilling

图 1-45 钻孔

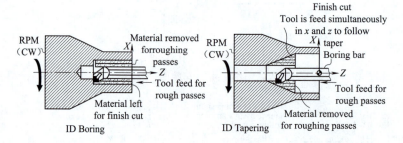

Figure 1-46 Boring

图 1-46 镗孔

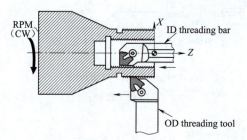

Figure 1-47 Threading

图 1-47 车螺纹

1. Facing

1. 车端面

This operation involves cutting the end of the stock such that the resulting end surface is perpendicular. A smooth, flat end surface should be produced. The tool is fed into the work in a direction perpendicular to the stock centerline.

该操作切削坯料的端部,使所得的端面垂直于坯料中心线。加工出光滑的端面。刀具应以垂直于中心线的方向切削工件。

2. Turning

2. 车外圆

Turning involves the removal of material from the outside of rotating stock. Rough turning is usually done first in multiple passes along the Z-axis. For each X-axis depth value, the tool removes stock as it travels along the Z-axis at feed rate. Different profile shapes can be created including tapers and contours.

车削从旋转坯料的外圆去除材料。通常在沿 Z 轴的多次车削中首先进行粗车。对于每次 X 轴的切削深度,刀具按进给速率沿 Z 轴行进时进行车削。可以形成不同的轮廓形状,包括锥形和轮廓。

In finish turning the servo motors move simultaneously in the X and Z directions at feed rate so that the finished shape is cut. The actual path followed by the tool is a series of straight line approximations generated by software algorithms.

在精加工中,伺服电动机以进给速率在 X 和 Z 方向上同时移动,从而切削零件形状。实际路径是由软件算法生成的一系列近似直线。

3. Chamfer

3. 倒角

A chamfer is a smooth beveled edge machined on the inside or outside of a finished diameter. A chamfer breaks a sharp edge and makes the finished part easier to assemble. Normally the chamfer angle is 45° with the chamfer distance between 0.25 to 6.35 mm. The servo motors move the tools simultaneously in the X and Z directions at feed rate to cut the chamfer.

倒角是在工件直径的内侧或外侧加工的光滑斜边。倒角去除了锋利的边缘,使零件更容易装配。通常倒角角度为45°,倒角距离在 0.25~6.35 mm 之间。伺服电动机以进给量同时在 X 轴和 Z 轴方向移动刀具,切削倒角。

4. Grooving

4. 车槽

Grooving requires that the tool be fed into the work in a direction perpendicular to the work's centerline along the X-axis. In face grooving, the tool is fed into the work in a direction parallel to the work's centerline along the Z-axis. The cutting edge of the tool is on its end. Grooving for thread relief is usually done prior to threading to ensure the resulting threads will be fully engaged up to the shoulder.

车槽要求刀具沿垂直于工件中心线的方向沿 X 轴对工件进行加工。在端面车槽中,刀具沿着 Z 轴平行于工件中心线的方向对工件进行加工。刀具的切削刃在其端面。通常在螺纹加工之前加工螺纹退刀槽，以确保最终的螺纹完全啮合到底部。

5. Parting

5. 切断

Parting involves cutting off the part from the main bar stock. Parting is done with a cutoff tool that is tapered and has a cutting edge at its end. The tool is fed into the part in a direction perpendicular to its centerline until the part is completely separated from the main bar stock.

切断是指将工件从棒料上分离。使用锥形切削刀具完成切削，切削工具刀口在其端面。刀具沿垂直于其中心线的方向进给，直到零件与棒料完全分离。

6. Drilling

6. 钻孔

The drill is usually mounted in a drill chuck or held in a bushing and fed into the rotating work along the Z-axis. A center drill should be applied before using a high speed steel twist drill, but it's not necessary if a spade drill or carbide insert drill is selected. Spade and carbide drills, however, need higher speed and a high pressure coolant system to operate.

钻头通常安装在钻头夹头中或固定在衬套中，并沿 Z 轴进给到旋转工件中。在使用高速钢麻花钻之前应使用中心钻，如果选择铲形钻或硬质合金钻头钻，则无须使用。然而，铲式和硬质合金钻头需要更高的转速和高压冷却系统才能使用。

7. Boring

7. 镗孔

Boring is an internal turning operation. Rough boring is usually done first in multiple passes along the Z-axis. For each X-axis depth value, the tool removes stock as it travels along the Z-axis at feed rate.

镗孔是一种内部车削操作。沿着 Z 轴进行多次切削时，首选进行粗镗。对于每个 X 轴向切削深度，刀具按进给量沿 Z 轴进行切削。

In finish tuning the servo motors move the tool simultaneously in the X and Z directions at feed rate in such a way that the finished shape is cut. The actual path followed by the tool is a series of straight line approximations generated by software algorithms. Boring can be used to more accurately size and true a hole, as well as create internal tapers and contours.

精镗时，伺服电动机以一定的进给量在 X 和 Z 方向上同时移动刀具，以便切削成最终的形状。该刀具实际路径是由软件算法生成的一系列近似直线。镗孔可用于更精确地获得孔的尺寸、内锥度和轮廓。

8. Threading

8. 车螺纹

This operation involves cutting helical grooves on the outside or inside surface of a cylinder or cone. Grooves or threads have a specific angle. Unified threads have an angle of 60°. The distance between teeth is called the thread pitch. The tool usually is fed into the material at a cut angle of 29° in a direction perpendicular to the part's centerline along the X-axis and at feed rate equal to the thread pitch.

该操作在圆柱体或圆锥体的外表面或内表面上切削出螺纹。螺纹具有特定角度,常见的螺纹角度为60°,齿之间的距离称为螺距。刀具通常沿着X轴垂直于毛坯中心线的方向以29°的切削角进给到材料中,并且进给速率等于螺距。

1.5 Tooling for CNC Lathe Operations
1.5 数控车床刀具

Modern CNC turning centers utilize tool holders with replaceable/indexable inserts. Tool holders come in a variety of styles. Each style is suited for a particular type of cutting operation as illustrated.

现代数控车削中心采用带可更换/可转位刀片的刀柄。刀柄有多种类型。每种类型都适用于特定的切削操作。

The advantages of using carbide insert tools, including cutting capabilities at higher speeds (two to three times faster than high-speed steel cutters) and reductions in tool inventory and elimination of regrinding time and cost. Another advantage of single-point tools is the fact that inserts are made with a precise tool nose radius for cutting, so the location of the tool nose center for any tool can be accurately determined. This makes tool offset specifications and CNC programming a much easier job. See Figure 1-48.

使用硬质合金刀片的优势是使用更高的切削速度(比高速钢刀具快2~3倍)和减少刀具库存以及节约时间和成本。另一个优点是,刀片是用一个固定的刀尖半径来切削,因此可以精确地确定刀具刀尖圆角圆心位置,这使得设置刀具补偿和数控编程更容易,如图1-48所示。

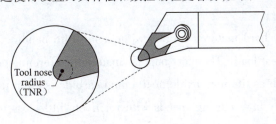

Figure 1-48 Tool Nose Radius
图1-48 刀尖圆弧半径

With regard to lathe operation, there are four important types of materials used for inserts.
对于车床操作,刀片有四种类型的重要材料。

视频
刀具安装方法3D视频

1. Cemented Carbides
1. 硬质合金

Cemented carbides are formed by using tungsten carbide sintered in a cobalt matrix. Some grades contain titanium carbide, tantalum carbide, or other materials as additives. Their chief advantage is the ability to cut at higher speeds and feeds. Carbide tools cut 5 to 10 times faster than high-speed steel tools. They offer excellent resistance to consistent high heat, thermal shock, impact, wear and abrasion. They also produce lower surface roughness for finishing cuts.

通过使用在钴基质中烧结的碳化钨形成硬质合金。某些牌号使用碳化钛、碳化钽或其他材料作为添加剂。它们的主要优势是能够以更高的切削速度和进给量进行切削。硬质合金刀具的切削速度比高速钢刀具快5~10倍。它们具有出色的耐高温、耐热冲击、耐冲击、耐用和耐磨的特性。精加工时,还能产生较好的表面粗糙度。

2. Coated Carbides
2. 涂层合金

The wear resistance of cemented carbides can be improved by 200% to 500% by coating them with wear-resistant materials. Coating materials include titanium carbide and aluminum oxide (a ceramic). The best resistance to abrasion wear for speeds below 500 m/min is achieved when titanium carbide is used. For higher speeds, resistance to the chemical reaction between the tool and work-piece is afforded by ceramic-coated inserts. Both coatings offer excellent performance on steels, cast iron, and nonferrous materials.

通过用耐磨材料涂层,硬质合金的耐磨性可提高200%~500%。涂层材料包括碳化钛和氧化铝(陶瓷)。当使用碳化钛时,对于低于500 m/min的速度,可以获得最佳的耐磨性。对于更高的速度,陶瓷涂层刀片可以减少刀具和工件之间的化学反应。两种涂料均可在钢、铸铁和有色金属材料上呈现出色的性能。

3. Ceramics
3. 陶瓷

A ceramic is a very hard material formed without metallic bonding. It displays exceptional resistance to wear and heat load. The most popular material for forming ceramics is aluminum oxide, but an additive such as titanium oxide or titanium carbide is often used. Hard materials can be machined at extremely high cutting speeds with relatively little loss in tool life. In addition, the surface finish is better than with other cutting materials. The main disadvantage with ceramics is that they have low resistance to impact and shock. Thus, they can only be used in operations where impact loading is low.

陶瓷是一种没有金属键合的硬质材料,并耐热和耐磨。氧化铝是一种常用于制造陶瓷的材料,而氧化钛和碳化钛常常作为添加剂。硬度高的材料可以在极高的切削速度下加工,而对刀具寿命损失相对较小。此外,精加工后表面粗糙度优于其他切削材料。陶瓷的主要缺点是具有低的抗冲击性。因此,它们只能用于冲击载荷较低的加工中。

4. Diamonds

4. 金刚石

There are two main types of diamond cutting materials. Single-crystal natural diamonds offer outstanding wear resistance but low shock resistance. Smaller synthetic diamond crystals fused together at high temperature and pressure into a carbide substrate form a material developed. It displays good resistance to shock loading. Diamond tools offer substantial improvements over carbides. Better surface finish and higher cutting speeds with substantial improvements in tool life can be achieved.

有两种主要类型的金刚石用于切削材料。单晶天然钻石具有出色的耐磨性,但抗冲击性低。较小的合成金刚石晶体在高温和高压下熔合在一起形成碳化物基材,具有良好的抗冲击负荷能力。金刚石刀具相比碳化物有了显著的改进,可以切削出更好的表面粗糙度,具有更高的切削速度,以及能够延长刀具寿命。

5. Insert Shape

5. 刀片形状

The seven basic shapes for turning operations in order of decreasing strength are shown in Figure 1-49 below.

按照强度逐渐降低的顺序,车削操作中使用的七种刀具形状如图 1-49 所示。

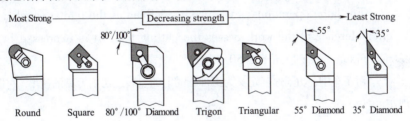

Figure 1-49　Various CNC Insert Shape

图 1-49　各类数控车刀

Some important rules to be considered when selecting tooling for the lathe operations are:

(1) Select an insert that has the highest strength shape possible.

(2) Select the smallest practical insert size (to save on cost).

(3) Select the largest tool nose radius (TNR) possible.

● A large TNR provides a strong cutting edge for roughing.

● A large TNR permits higher feed rates for roughing.

● Select a smaller TNR if the operation tends to produce vibration during roughing.

● For roughing, the most commonly used TNR values are: 0.4 mm to 1.2 mm.

(4) Select the largest boring bar diameter with the smallest possible overhang to reduce de-

flection and vibration. A solid steel shank boring bar can normally have an overhang of up to four times the diameter.

选择车刀时,要考虑的一些重要原则是:
(1)选择具有最高强度形状的刀片。
(2)选择最小的刀片规格(以节省成本)。
(3)尽可能选择最大刀尖圆弧半径(TNR)。
- 大的 TNR 为粗加工提供了高强度的切削刃。
- 较大的 TNR 允许较高的粗加工进给量。
- 如果在粗加工操作过程中容易产生振动,请选择较小的 TNR。
- 对于粗加工,最常用的 TNR 值为 0.4~1.2 mm。

(4)选择最大的镗杆直径和尽可能小的伸缩量,以减少变形和振动。实心钢柄镗杆通常可以具有高达直径四倍的伸缩量。

1.6　Tool Speeds, Feeds, and Depth of Cut for Lathe Operations

1.6　数控车床的切削速度、进给量及切削深度

视频
外圆车刀的三面二刃一尖

1. Tool Speed
1. 切削速度

For lathe operations, tool speed is defined as the rate at which a point on the circumference of the work passes the cutting tool. It is expressed in surface feet per minute. See Figure 1-50.

对于车削,切削速度定义为工件圆周上的点通过切削刀具的速率。它以英寸每分表示,如图1-50所示。

In English units:

$$\text{Spindle}(\text{r/min}) = \frac{4 \times \text{Tool Cutting Speed}(\text{in/min})}{\text{Cut Diameter}(\text{in})}$$

以英制单位,主轴转速的计算公式:

$$\text{主轴转速}(\text{r/min}) = \frac{4 \times \text{切削速度}(\text{in/min})}{\text{切削直径}(\text{in})}$$

In metric units:

$$\text{Spindle}(\text{r/min}) = \frac{1\,000 \times \text{Tool Cutting Speed}(\text{m/min})}{\pi \times \text{Cut Diameter}(\text{mm})}$$

以米制单位,主轴转速的计算公式:

$$\text{主轴转速}(\text{r/min}) = \frac{1\,000 \times \text{切削速度}(\text{m/min})}{\pi \times \text{切削直径}(\text{mm})}$$

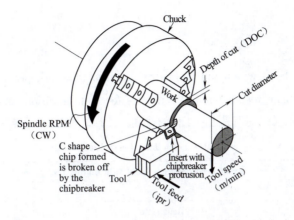

Figure 1-50 Tool Speed

图 1-50 刀具速度

The approximate recommended cutting speeds for high-speed steel cutting tools are listed in Table 1-2.

表 1-2 列出了高速钢刀具的推荐切削速度。

The values for tool speed listed in Table 1-2 can be doubled for carbide-tipped cutting tools.

对于硬质合金刀具,刀具速度是表 1-2 中数值的两倍。

Table 1-2 Cutting Speed

Approximate turning speeds for high-speed steel cutting tools. Depth of rough cut is between 3 and 6 mm		
		Tool speed(m/min)
Material	Rough cuts	Finish cuts
Aluminum	400	800
Brass	250	500
Cast iron	75	200
Mild steel	100	250
Tool steel	50	100

表 1-2 切削速度

适用于高速钢车刀。粗加工的切削深度在 3~6 mm		
		切削速度(m/min)
材料	粗加工	精加工
铝	400	800
黄铜	250	500
铸铁	75	200
低碳钢	100	250
工具钢	50	100

2. Tool Feed
2. 刀具进给量

Tool feed is the rate at which the tool advances into the work per revolution of the work. Table 1-3 lists approximate recommended feed rates for cutting various materials.

刀具进给量是工件旋转一周刀具进入工件的速度。表1-3列出了切削各种材料的推荐进给量。

Table 1-3　Tool Feed Speed

Turning Feeds for High-Speed Steel Cutting Tools		
		Tool feed(mm/r)
Material	Rough cuts	Finish cuts
Aluminum	0.38~0.64	0.13~0.25
Brass	0.38~0.64	0.08~0.25
Cast iron	0.38~0.64	0.13~0.25
Mild steel	0.25~0.5	0.08~0.25
Tool steel	0.25~0.5	0.08~0.25

表1-3　刀具进给量

高速钢刀具的进给量		
		刀具进给量(mm/r)
材料	粗加工	精加工
铝	0.38~0.64	0.13~0.25
黄铜	0.38~0.64	0.08~0.25
铸铁	0.38~0.64	0.13~0.25
低碳钢	0.25~0.5	0.08~0.25
工具钢	0.25~0.5	0.08~0.25

3. Depth of Cut（DOC）
3. 切削深度（DOC）

The depth of cut is the thickness of the material machined from the work or is the distance from the uncut work surface to the cut surface. The diameter of the work remaining after the Cut is given as:

$$\text{Cut diameter} = \text{Uncut Diameter} - 2 * \text{DOC}$$

切削深度是从工件上切削下来的材料的厚度，或者是从未切削的工作表面到切削表面的距离。切削后剩余工件的直径如下：

$$\text{切削后直径} = \text{未切削直径} - 2 \times \text{切削深度}$$

In machining a part, rough cuts are taken first. For this operation the DOC is taken as large as possible with the feed reduced. Factors that influence DOC in roughing are the power of the CNC, the material and shape of the insert, the rigidity of the work, and the feed.

在加工零件时，首先进行粗加工。对于该操作选择尽可能大的切削深度以减少切削次数。影响粗加工中切削深度的因素是数控机床的功率、刀片的材料和形状、工件的刚性和进给量。

For the one or two finishing cuts that follow roughing, a DOC is taken relatively lightly with the feed adjusted to produce a required surface finish.

对于粗加工后的一次或两次精加工切削,切削深度较小,并调整进给量加工出所需的表面粗糙度。

Below are listed recommended DOC values for insert tool nose radii when executing finishing passes, see Table 1-4.

视频

数控车退刀方式

下面列出了精加工过程中刀片推荐切削深度,见表1-4。

Table 1-4　Recommended DOC values

Tool Nose Radius of Insert(mm)	Recommend DOC for Finishing Passes(mm)
0.38	0.12
0.79	0.23
1.22	0.35
1.60	0.46
2.39	0.70
3.18	0.94

表1-4　推荐DOC值

刀片的刀尖圆弧半径(mm)	推荐用于精加工的切削深度(mm)
0.38	0.12
0.79	0.23
1.22	0.35
1.60	0.46
2.39	0.70
3.18	0.94

1.7　Feed Directions and Rake Angle for Lathe Operations

1.7　数控车床操作过程中的进给方向及前角

1. Feed Directions

1. 进给方向

Right-handed cutting tools have their cutting edge on the right side and are usually fed from right to left. Left-handed cutting tools have the cutting edge on the left side and are fed left to right. A neutral cutting tool has cutting edges on both the right and left sides and can be fed right to left or left to right.

It is used for profiling where clearance on both sides is required. See Figure 1-51.

右偏刀的切削刃在刀具右边,通常从右向左进给切削。左偏刀的切削刃在左边,从左到右进给切削。中性刀在左右两侧都有切削刃,可以从右到左或从左到右进给切削,它用于切削两侧都需加工的零件,如图 1-51 所示。

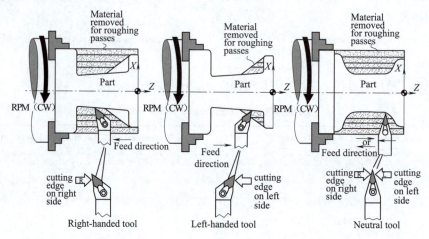

Figure 1-51   The Right-hand Rule

图 1-51   右手法则

2. Tooling Nomenclature
2. 刀具命名

Single-point turning and facing tools must have specific angles at their cutting ends to ensure longer tool life and cutting efficiency, as shown in Figure 1-52.

单点车刀和端面车刀必须在其切削刃具有特定角度,以确保更高的刀具寿命和切削效率,如图 1-52 所示。

The components of Figure 1-52 are as follows:
图 1-52 的组成如下:

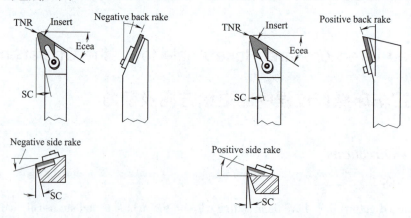

Figure 1-52   Single Point Tool Angles (Angles Shown Exaggerated for Emphasis)

图 1-52   单点切削刀具的角度(为了强调而放大显示的角度)

TNR: Tool nose radius for producing acceptable surface finish and longer tool life.

刀尖圆弧半径:可产生所需的表面粗糙度并延长刀具寿命。

BR: Back rake angle for directing the chips away from the work and toward the tool holder. This is the angle made between the top face of the carbide insert and the tool shank in the length direction.

后倾角:控制切屑流向,使其从工件流向刀杆。这是硬质合金刀片的顶面与刀柄在长度方向上形成的角度。

SR: Side rake angle for directing chips away from the work and toward the side. This is the angle made between the top face of the carbide insert and the tool shank in the width direction.

侧倾角:用于引导切屑废料离开工件并朝向侧面。这是硬质合金刀片的顶面与刀柄在宽度方向上形成的角度。

SC: Side clearance angle for permitting the side of the tool to enter the work.

侧后角:允许刀具侧面进入工件。

EC: End clearance angle for permitting the end of the tool to enter the work.

前端后角:允许刀具末端进入工件的端部间隙角。

SCEA: Side cutting edge angle for improving the shear cut and producing thinner chips when turning.

侧切削刃角:用于改善切削并在车削时产生更薄的切屑废料。

ECEA: End cutting edge angle for maintaining clearance between the tool and the work during boring or facing operations.

副偏角:在钻孔或端面切削操作期间,用于保持刀具与工件之间间隙。

3. Rake Angles and Cutting Force

3. 前角和切削力

As was stated previously, a tool's rake angles help carry away chips. These angles also help protect the cutting tool from excessive heating and abrasive action. In general, positive rake angles tend to decrease the force with which the tool cuts and negative rake angles increase the force. The decision to use a positive or negative rake tool depends upon many factors, chief among them is the material to be machined. A summary of the properties and use of negative verses positive rake tools is given in Table 1-5.

如前所述,刀具的前角有助于带走切屑废料。这些角度还有助于保护切削刀具免受过热和磨蚀作用。通常来说,正前角倾向于减小工具切削的力,而负前角反之。使用正或负前角刀具取决于许多因素,其中主要考虑被加工的材料。表1-5 总结了正前角和负前角的用法。

It is also best to select the tool holder style that creates the largest possible side cutting edge angle with respect to the work in order to produce thinner chips at lower cutting edge temperatures and thus protect the tool nose from excessive wear.

同样,最好选择相对于工件产生最大的侧切削刃角的刀杆样式,以便在较低的切削刃温度下生产更薄的切屑废料,从而保护刀尖免受过度磨损。

Table 1-5　The Properties and Use of Negative Verses Positive Rake Tools

Positive Rake Tools	Negative Rake Tools
Decreases the strength of the cutting edges	Increase the strength of the cutting edges
Machines with lower cutting force	Machines with higher cutting force
Lower power machining	Rigid, high power machining
Only top cutting edges can be used	More economical—both top and bottom cutting edges can be used
Use on aluminum, titanium, copper, most stainless steels, thin or easily deflected parts, nickel alloys and plastic	Use on cast iron, carbon steels, and for heavy, interrupted cuts

表1-5　正前角和负前角刀具的性质和用法

正前角刀具	负前角刀具
降低切削刃的强度	增加切削刃的强度
切削力较低的机床	切削力更高的机床
低功率加工	刚性,高功率加工
只能使用顶端切削刃	更经济——可以使用顶端和底端切削刃
用于铝、钛、铜、大多数不锈钢,薄或易偏转的零件,镍合金和塑料	用于铸铁、碳钢和重型间断切削

Tool manufacturers use variations of a standard coding system to specify the characteristics of their tool holders and boring bars. A coding system from Iscar Metals, Inc, is shown in Figure 1-53.

刀具制造商使用标准编码系统的变体来凸显其刀杆和镗杆的特性。伊斯卡金属公司的编码系统如图1-53所示。

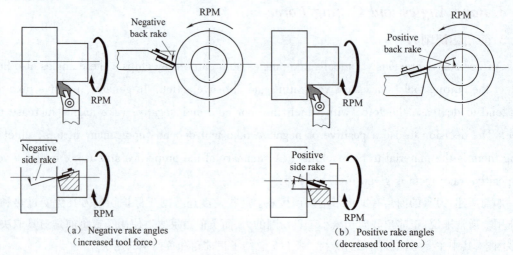

Figure 1-53　Rake Angles for Turning (angles shown exaggerated for emphasis)

图1-53　车削时的前角(为了强调而放大显示的角度)

Chapter Summary

本章总结

The following key concepts were discussed in this chapter:

(1) A CNC lathe is a machine designed to remove material from stock that is clamped and rotated.

(2) Most CNC lathes use turrets for holding, changing, and moving tools.

(3) The CNC lathe machine axes of motion are: $-z$ along the spindle axis and toward the spindle, $+z$ along the spindle axis away from the spindle, $-x$ transverse to the spindle axis and toward the spindle, and $+x$ transverse to the spindle axis and away from the spindle.

(4) The seven most important lathe operations are facing, turning, grooving, parting, drilling, boring, and threading.

(5) Modern CNC lathes and turning centers utilize indexable insert cutting tools.

(6) Single-point turning and facing tools have their cutting edges set at specific cutting angles to ensure longer tool life and greater cutting efficiency.

(7) Tool speed for lathes is the rate at which a point on the revolving circumference of the work passes the cutting tool. It is expressed in surface meter per minute (m/min).

(8) Tool feed is the rate at which the tool advances into the work per revolution of the spindle.

(9) Depth of cut is the thickness of the material removed by the tool or the Radial distance in inches from the uncut surface of the stock to the cut surface.

(10) Right-handed cutting tools have their cutting edges on the right side and are fed into the work from right to left. Left-handed tools have their cutting edges on the left side and are fed into the work from left to right.

(11) Negative rake angles are used for most turning operations involving carbide insert cutting tools. Positive rake angles tend to decrease tool force and are used with high-speed steel cutters on softer materials.

本章讨论了以下概念：

(1)数控车床是一种用于从旋转的材料中切除材料的机器。

(2)大多数数控车床使用刀塔来夹持、更换和移动刀具。

(3)数控车床的运动轴为：$-z$沿主轴轴线朝向主轴，$+z$沿主轴轴线远离主轴，$-x$平行于主轴轴线并朝向主轴，$+x$平行于主轴轴线远离主轴。

(4)七个最重要的车床操作是车端面、车外圆、车槽、切断、钻孔、镗孔和车螺纹。

(5)现代数控车床和车削中心使用可转位刀片刀具。

(6)尖点车刀和端面车刀的切削刃有特定的切削角度，以确保更长的刀具寿命和更高的切削效率。

（7）车床的切削速度指的是工件旋转圆周上的一个点通过刀具的速度。它以米每分（m/min）表示。

（8）刀具进给量是工件旋转一周刀具进入工件的距离。

（9）切削深度是刀具切削材料的厚度或从原料未切削表面到切削表面的径向距离。

（10）右偏刀切削刀具的切削刃在刀具的右侧，从右到左切入工件。左偏刀的左侧切削刃从左到右进给切削。

（11）负前角用于大多数涉及硬质合金刀片刀具的车削操作。正前角倾向于降低切削力，可采用高速钢刀切削较软的材料。

Review Exercises

回顾练习

1.1 Identify and define the components of the CNC turning center (Figure 1-54) and write the answer in Table 1-6.

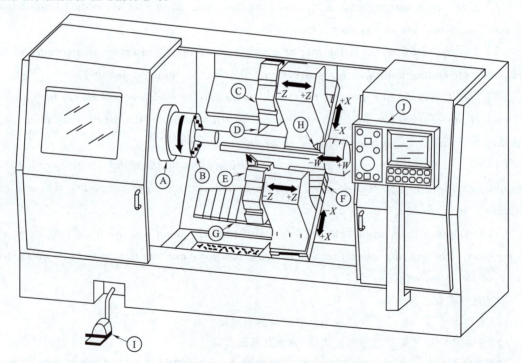

Figure 1-54 The CNC Turning Center
图 1-54 数控车削中心

Table 1-6 Identification and definition of the components of the CNC turning center

Component	Description/Function
A	
B	
C	
D	
E	
F	
G	
H	
I	
J	

1.1　识别和定义数控车削中心的组成部分(图1-54)，并将答案写在表1-6中。

表1-6　数控车削中心组成部分的定义和描述

组成部分	描述/功能
A	
B	
C	
D	
E	
F	
G	
H	
I	
J	

1.2 a. What type of tool changer mechanisms are used on CNC lathes and turning-centers?

b. Describe a typical tool changing operation.

1.2　a.在数控车床上和车削中心使用什么类型的换刀机构？

b.描述一个典型的换刀操作。

1.3 What is the difference between a front turret and a rear turret?

1.3　前置刀架和后置刀架的区别是什么？

1.4 Describe the CNC lathe machine axis of motion.

1.4　描述数控车床运动轴。

1.5 What are OD versus ID operations?

1.5　外圆车削与内孔车削操作是什么？

1.6 Describe the following machining operations and state whether each is OD, ID,

or both：

    a. Facing   b. Turning   c. Grooving   d. Parting   e. Boring   f. Threading

1.6   描述下面哪些加工内容是外圆车削、哪些是内孔加工，或者既是外圆加工也是内孔加工。

    a.车端面   b.车外圆   c.切槽   d.切断   e.镗孔   f.车螺纹

1.7   List four advantages of using indexable/insert tooling for lathe operations.

1.7   列出使用可转位/刀片车刀进行车削的四个优点。

1.8   State the advantages and disadvantages of using the following insert materials：

    a. Cemented carbides   b. Coated carbides   c. Ceramics   d. Diamonds

1.8   说明使用下列刀片材料的优点和缺点：

    a.硬质合金   b.涂层碳化物   c.陶瓷   d.金刚石

1.9   List four important rules that should be followed when selecting insert tooling.

1.9   列出在选择刀片时应遵循的四条重要原则。

1.10   Explain the following terms as applied to lathe operations：

    a. Tool speed   b. Tool feed   c. Depth of cut

1.10   解释下列车床操作的术语：

    a.切削速度   b.进给量   c.切削深度

1.11   Explain the difference between right-hand and left-hand cutting tools and how they are used in OD turning operations.

1.11   解释右偏刀和左偏刀之间的差异以及它们在外圆车削操作中的应用。

1.12   Match the terms on the left with the definitions on the right：

Side rake angle	Permits side of tool to enter work
Tool nose radius	Clearance for boring/facing
Side clearance angle	Directs chips toward the side
Side cutting edge angle	Produces surface finish
Back rake angle	Controls chip thickness
End cutting edge angle	Directs chips away from work

1.12   匹配左边的术语和右边的定义：

侧倾角	允许刀具刃口切削工件
刀尖圆弧半径	镗孔/车端面的间隙
侧后角	控制切屑流向侧面
侧切削刃角	产生最终的表面粗糙度
后倾角	控制切屑厚度
副偏角	控制切屑远离工件

1.13   a. What is the effect of using negative rake angle tooling？

    b. What is the effect of using positive rake angle tooling？

1.13   a.使用负前角刀具的效果是什么？

    b.使用正前角刀具的效果是什么？

1.14 What recommendations can be made regarding the rake angle when selecting a tool holder for the following machining situations?

a. Rough turning cast-iron blank

b. Rough turning an aluminum steel shaft

c. Rough turning a part made of carbon steel

d. Taking heavy interrupted cuts on stainless steel

1.14 对于在下列加工情况下选择刀杆时的前角可提出什么建议?

a. 粗加工铸铁件

b. 粗加工铝钢件

c. 粗加工碳钢件

d. 对不锈钢进行断续切削

Chapter 2 Fundamental Concepts of CNC Lathe Programming

第 2 章 数控车床编程的基本概念

Learning Objectives

At the conclusion of this chapter you will be able to:

(1) Explain the different types of positioning modes for CNC lathe operations.

(2) Know the important locations for programming and setup of CNC lathes. These include the reference point, machining origin, and program origin.

(3) Understand methodizing of operations for CNC lathes.

(4) State the basic setup operations performed on the CNC lathe prior to running a job.

(5) Know the important preparatory (G) codes and miscellaneous (M) codes used in programming lathe operations.

(6) State the codes for specifying values of the spindle speed, tool feed, and changing tools.

(7) Explain tool nose radius(TNR) compensation.

(8) Apply TNR compensation in programming lathe operations.

学习目标

在本章结束时,你将能够:

(1)解释数控车床操作不同类型的定位方式。

(2)知道数控车床编程和操作时重要的位置点。包括机床参考点、机床原点和编程原点。

(3)了解数控车床的操作方法。

(4)说明在加工前数控车床上执行的基本操作。

(5)了解车床加工编程中的准备功能(G)代码和辅助功能(M)代码。

(6)说明主轴转速、刀具进给量和更换刀具的代码。

(7)解释刀尖圆弧半径(TNR)补偿。

(8)在车床编程中应用刀尖圆弧半径补偿。

This chapter deals with the basic parameters and concepts involved in programming CNC lathe operations. See Figure 2-1. Modes for positioning the tool are considered first. Important setup information concerning the reference.

本章论述了数控车床操作编程中涉及的基本参数和概念。如图 2-1 所示,首先考虑刀具的定位方式以及相关重要的操作信息。

Chapter 2 Fundamental Concepts of CNC Lathe Programming
第 2 章　数控车床编程的基本概念

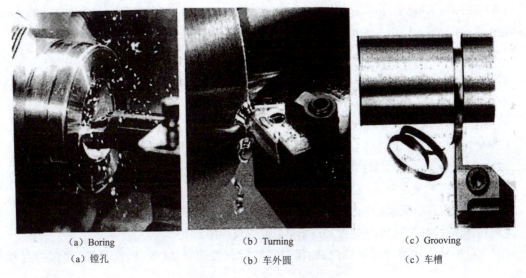

(a) Boring　　　　　　　　(b) Turning　　　　　　　　(c) Grooving
(a) 镗孔　　　　　　　　　(b) 车外圆　　　　　　　　(c) 车槽

Figure 2-1　CNC Lathe Operations
图 2-1　数控车床的操作

Part origin, tool change point, and tool offsets are explored in detail. The planning of machining operations for CNC lathes is discussed. A sample of setup and tool and operations sheets for a lathe job is presented. Important preparatory (G) codes and miscellaneous (M) codes for executing basic lathe operations are listed and explained. Additionally, feed rate (F) codes, speed (S) codes, and tool change (T) codes are also studied.

本章详细地说明了编程原点、刀具换刀点和刀具补偿。探讨了如何规划数控车床的加工操作。给出了工件的安装和刀具操作的示例。列出和说明了执行基本车床操作的重要准备功能(G)代码和辅助功能(M)代码。此外，还说明了进给量(F)代码、主轴速度(S)代码和换刀(T)代码。

Because the tool nose radius is so small, tool movements are not normally programmed from the center of the tool nose. Instead, programmers use tool nose radius (TNR) compensation, which allows tool movements to be programmed directly from the part geometry. Important concepts of TNR, as well as detailed TNR programming examples, are featured.

由于刀尖圆弧半径太小，刀具运动通常不能从刀尖的圆心编程。相反，程序员使用刀尖圆弧半径(TNR)补偿，这允许刀具运动直接从零件几何形状编程。本章论述了刀尖圆弧半径的概念，并给出详细的刀尖圆弧半径编程示例。

When Job setup on a CNC turning center, safety rules are:

(1) Hold the work piece securely. For bar work, make sure the bar does not extend beyond safe limits.

(2) Use safety goggles, wear protective clothing, and put long hair up.

(3) Set the jaw pressure to conform to proper holding conditions.

(4) Set cutting speeds and feeds for each tool within the limits recommended for the process.

(5) Adjust speeds and feeds during an operation to obtain optimum machining conditions.

(6) When moving an axis in manual mode, make sure there is sufficient clearance between the cutting tool and all surrounding objects (part, fixture, etc.).

(7) Test the program by making a dry run.

(8) When loading, make sure the workpiece is free of burrs and foreign particles.

(9) Backup setup and the program file.

当执行车削操作时,安全规程是:

(1)安全装夹工件。加工棒料时,确保棒料长度不超出安全界限。

(2)使用安全护目镜,穿防护服,盘起长发。

(3)设定合适的装夹力,保证夹持稳固。

(4)在工艺推荐的限制范围内设定每个刀具的切削速度和进给量。

(5)在操作过程中调整切削速度和进给量以获得最佳加工条件。

(6)当以手动方式移动运动轴时,确保刀具与所有周围物体(零件、夹具等)之间有足够的间隙。

(7)通过试运行测试程序。

(8)当安装毛坯时,确保工件没有毛刺和异物。

(9)备份安装程序和程序文件。

2.1　Establishing Locations via Cartesian Coordinates

2.1　通过笛卡儿法则建立坐标系

The machine axes of motion for CNC lathes were introduced. It was pointed out that for turning centers, the X axis (cross-slide movement) and the Z axis (longitudinal travel) are used to specify tool locations. See Figure 2-2.

本节介绍了数控车床的运动轴。对于车削中心,X轴(横向移动)和Z轴(纵向行程)用于指定刀具位置,如图2-2所示。

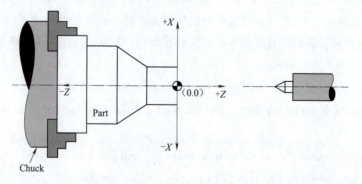

Figure 2-2　The Cartesian Coordinate System for Rear Turret CNC Lathes

图2-2　后置刀架数控车床的笛卡儿坐标系

2.2　Types of Tool Positioning Modes

2.2　刀具定位方式

CNC lathe programs can be written to move the tool in the following modes: absolute, incremental, or mixed (incremental and absolute).

数控车床程序可以按照绝对、增量或混合(增量和绝对)方式移动刀具。

1. Absolute Positioning

1. 绝对坐标编程

视频
机床坐标系

When operating in this mode the new position of the tool is given by its X and Z distances from a fixed home or origin (0, 0).

在此模式下操作时,点的位置由 X 和 Z 方向到坐标原点(0,0)的距离衡量。

With absolute diameter programming, the X position of the tool is specified in terms of diameter or twice the distance from the spindle centerline. Refer to Figure 2-3 and Table 2-1.

使用绝对坐标编程时,刀具的 X 向位置根据直径或距主轴中心线的距离的两倍来指定,见图2-3和表2-1。

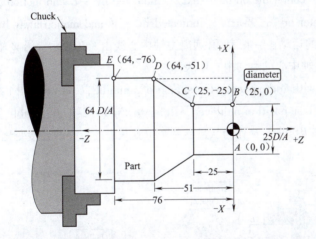

视频
正交平面参考系

Figure 2-3　Programming by diameter (absolute positioning)

图 2-3　按直径编程(绝对坐标)

Table 2-1　Absolute Positioning

Tool position	Location (absolute)	
	X [diameter]	Z
A	0	0
B	25	0
C	25	−25
D	64	−51
E	64	−76

表 2-1 绝对坐标

刀具位置	位置(绝对值)	
	X [直径]	Z
A	0	0
B	25	0
C	25	-25
D	64	-51
E	64	-76

2. Incremental Programming

2. 增量坐标编程

When operating in this mode of programming, the new position of the tool is specified by inputting its direction and distance from the last position achieved. The address U is used to indicate incremental X-axis motion and the address W is used to indicate incremental Z-axis motion.

在这种程序模块下操作时,刀具位置通过输入其最终位置和前一点位置的矢量量值来确定。地址 U 表示 X 轴增量,地址 W 表示 Z 轴增量。

Motion toward the spindle centerline on the X-axis is indicated by $-U$ and motion away by $+U$. Motion toward the spindle center on the Z-axis is indicated by $-W$ and motion away by $+W$.

沿 X 轴向的主轴中心线的运动由 $-U$ 表示,远离主轴中心线由 $+U$ 表示。沿 Z 轴朝向主轴中心的运动由 $-W$ 表示,远离主轴中心的运动由 $+W$ 表示。

Diameter programming with incremental diameter programming, U is entered as the directed change in diameter from the last position achieved. Refer to Figure 2-4 and Table 2-2.

直径编程使用增量坐标编程,输入 U 作为从最后一个位置获得的直径的直接变化,见图 2-4 和表 2-2。

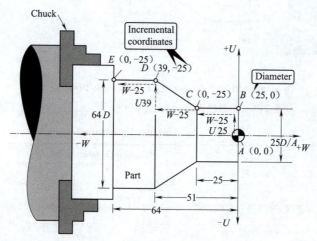

Figure 2-4　Programming by diameter (incremental positioning)

图 2-4　按直径编程(增量坐标)

Table 2-2   Incremental Positioning

Tool position	Location (incremental)	
	U	W
A	0	0
B	25	0
C	0	−25
D	39	−25
E	0	−25

表 2-2   增量坐标

刀具位置	位置(增量值)	
	U	W
A	0	0
B	25	0
C	0	−25
D	39	−25
E	0	−25

2.3　Reference Point, Machine Origin, Program Origin (FANUC Controllers)

2.3　机床参考点、机床原点和编程原点（FANUC 控制器）

There are three important feature locations to consider for FANUC system CNC lathes. See Figure 2-5.

FANUC 系统的数控车床有三个重要的特征位置需要考虑，如图 2-5 所示。

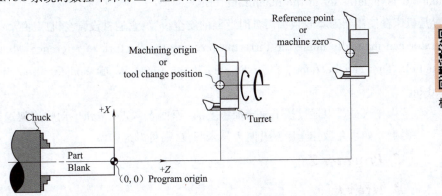

Figure 2-5   Important Locations for CNC Lathes

图 2-5　CNC 车床的重要位置点

1. Reference Point (Machine Zero)
1. 机床参考点(机床零点)

The reference point or machine zero is the position of the turret when the machine's axes are zeroed out. It is set once by the manufacture. It should be noted that this location in also referred to as the "machine home" position.

机床参考点或机床零点是机床轴被归零时刀架的位置。它由制造商设定,应该指出的是,这个位置也被称为"机床初始点"的位置。

2. Machining Origin (Tool Change Position)
2. 机床原点(换刀位置)

The machining origin is determined at setup. This location is input at the beginning of the word address program before the first tool change block by means of the "zero-offset" command. The machine indexes the tools at this position.

机床操作开始时确定机床原点。通过"零偏移"命令,在第一个换刀之前的字地址程序的开头输入该位置,机床在此位置替换刀具。

机床原点

The machining origin is determined in such a way that when the turret is at this location the longest tool is at least 1.0 from the face of the part in the "Z"-direction and 1.0 from the diameter of the stock. The control is zeroed out at this location. The turret is then jogged to the reference point. The "X" and "Z" locations of the machine origin from the reference point are recorded, and these numbers are used in the "zero-offset" command.

机床原点的确定方式:当刀架处于该位置时,最长的刀具距离部件的"Z"方向的面至少为1.0,距离坯料的直径至少为1.0。机床在此位置归零,然后将刀架点动移动到机床参考点。记录来自机床参考点的机床原点的"X"和"Z"位置,并且这些数字用于"零偏移"命令。

The tool change position is a safe location the machine returns to when indexing an old tool with a new tool. It is usually set at the machining origin.

换刀位置是机床在使用新刀具替换旧刀具时返回的安全位置,它通常设置在机床原点。

The operator can manually home the turret by pressing the "return to reference" button on the machine panel. This can be done, for example, if the turret is not homed when the CNC lathe is turned on.

编程坐标系

操作人员可以通过按下机床面板上的"返回参考点"按钮手动将刀架返回机床参考点。如果在数控车床打开时刀架未归位,则可以这样做。

3. Program Zero
3. 编程零点

Program zero is a point from which all dimensions are defined in the part program. The setup person uses tool offsets as a means of locating the program zero with respect to the machining origin.

编程零点是零件程序中定义所有尺寸位置的基准点。操作人员使用刀具补偿作为相对于机床原点定位编程零点的方法。

Upon receiving a programmed X and Z move with respect to the program zero, the controller will compute the corresponding X and Z move relative to the machining origin. It will then execute the move relative to the machining origin.

在接收到相对于编程零点的编程的 X 轴和 Z 轴的移动命令时,控制器将计算相对于机床原点的相应 X 轴和 Z 轴移动,然后将执行相对于机床原点的移动。

2.4 Methodizing of Operations for CNC Lathe
2.4 数控车床的操作方法

Methodizing for lathes and turning centers carries the same meaning as described for machining centers. It is a plan that specifies the sequence and methods in which operations are to be carried out in order to produce a part. In a small shop, methodizing is done by the programmer, but in larger shops manufacturing engineers take on this job.

车床和车削中心的加工方法与加工中心的加工方法具有相同内容,指定了加工生产零件的顺序和方法。在小厂家中,方法由程序员完成,但在较大的厂家中,制造工程师承担这项工作。

1. Deciding on a CNC Lathe
1. 数控车床的选定

The programmer selects a CNC lathe based on its ability to optimize the cut-ting operations required to produce the part. The machine must have adequate travel, horsepower, turret size, accuracy and number of machine axis movements.

程序员根据能优化生产零件所需的切削操作选择数控车床。机器必须具有足够的行程、马力、刀架尺寸、精度和运动轴数。

2. Methods of Holding the Part During Machining
2. 加工过程中零件的装夹方法

The method of holding the part during machining is determined by the programmer. The set-up person acts on a sketch provided by the programmer, For simple bar shapes a hydraulic chuck is used; for more complex shapes a faceplate and fixture is used. All necessary arresting and support devices such as clamps, counterweights, angle plates and their locations must be carefully documented. Refer to Figure 2-6.

在加工过程中如何装夹零件是由程序员决定的。操作工根据程序员提供的图纸进行操作。对于简单的棒状工件,使用液压卡盘;对于复杂的工件,使用面板和夹具。必须仔细检查所有夹具,如夹具、配重、角板等,如图2-6所示。

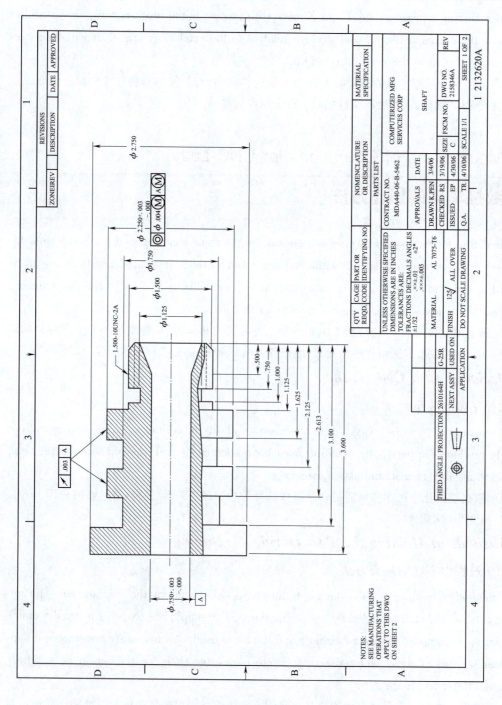

Figure 2-6　A Part Production Drawing for Manufacturing the Part SHAFT

图 2-6　用于制造零件的零件加工图

3. Machining Determination

3. 数控加工方案的确定

After making a thorough study of the part print, the programmer plans the machining sequence and the corresponding tooling required.

在对零件图进行彻底研究之后,程序员开始设计加工顺序,准备所需的刀具。

In general, machining on CNC lathes should proceed in the following order:

(1) Rough and finish facing.

(2) Rough and finish turning.

(3) Drilling.

(4) Rough and finish boring.

(5) Grooving.

(6) Threading.

(7) Parting.

通常,数控车床上的加工应按以下顺序进行:

(1) 粗加工和精加工端面。

(2) 粗加工和精加工外圆。

(3) 钻孔。

(4) 粗镗和精镗。

(5) 车槽。

(6) 车螺纹。

(7) 切断。

The cutting sequence together with the required cutting tools is documented on the CNC Tool and Operations sheet. See Table 2-3.

数控加工和刀具表中记录了加工顺序以及所需的切削工具,见表2-3。

Table 2-3 The CNC Tool and Operations Sheet for Machining for Part SHAFT

COMPUTERIZED MFG SERVICE CORP CNC TOOL AND OPERATIONS SHEET						
BLUEPRINT NO:2158346A SHEET:1 OF 1							
JOB NO:6347-18 MATERISL: AL ALLOY 7075-T6 PROGRAM NO:O2000 MACHINE:KT 15/21 KIA LATHE			PART NO:84-764522-6 PART NAME:SHAFT PREPARED BY:M. CHIPS DATE:6/27/06			
TOOL	OFF-SET	OPERATION	TOOL DESCRIPTION	INSET	TNR
1	01	FACE END	ISCAR TURNING GOLDER MCLNR/L	CNMG-432	0.8

Continued

TOOL	OFF-SET	OPERATION	TOOL DESCRIPTION	INSET	TNR
2	02	ROUGH TURN OD LEAVE 0.25 mm FOR FINISHING	ISCAR TURNING GOLDER MCLNR/L 12-4	CNMG-432-NF	0.8
3	03	FINISH OD COUNTOUR	ISCAR TURNING GOLDER MCLNR/L 12-4	CNMG-432-NR	0.8
4	04	GROOVE OD	ISCAR TURNING GOLDER MCLNR/L 25.4-5	DGN 5003Z	0.3
5	05	CUT 1.50-10UNC-2A THREAD	ISCAR TURNING GOLDER MCLNR/L 0705K15	16 ER 10	0.3
6	06	CENTER DRILL X 4.5 mm DEEP	VALENITE .25DIA CENTER DRILL		
7	07	LATHE DRILL THRU	VALENITE .75DIA CENTER DRILL		
8	08	ROUGH BORE IDLEAVE 0.25 mm FOR FINISHING	ISCAR TURNING GOLDER MCLNR/L 20-4	CNMG-432-NR	0.8
9	09	FINISH ID	ISCAR TURNING GOLDER MCLNR/L 20-4	CNMG-432-NF	0.8
10	10	LATHE CUTOFF	ISCAR TURNING GOLDER MCLNR/L 19-4	GFN 6	0.35

表2-3 用于零件轴加工的CNC刀具和操作表

MFG计算机服务公司
数控加工和刀具表

图纸编号:2158346A
图纸:1号

工作号:6347-18　　　　　　　　产品编号:84-764522-6
材料:铝合金7075-T6　　　　　　零件名称:轴
程序编号:O2000　　　　　　　　准备:M.CHIPS
机床:KT 15/21 KIA数控车床　　　日期:2006年6月27日

刀具号	补偿号	操作	刀具说明	刀片规格
1	01	车端面	ISCAR TURNING GOLDER MCLNR/L	CNMG-432
2	02	粗车外圆,留0.25 mm的精加工余量	ISCAR TURNING GOLDER MCLNR/L 12-4	CNMG-432-NF
3	03	精车外圆	ISCAR TURNING GOLDER MCLNR/L 12-4	CNMG-432-NR
4	04	车槽	ISCAR TURNING GOLDER MCLNR/L 25.4-5	DGN 5003Z

续表

刀具号	补偿号	操作	刀具说明	刀片规格
5	05	车螺纹 1.50-10UNC-2A	ISCAR TURNING GOLDER MCLNR/L 0705K15	16 ER 10
6	06	钻中心孔，X 方向 4.5 mm 深	VALENITE .25DIA CENTER DRILL	
7	07	钻孔	VALENITE .75DIA CENTER DRILL	
8	08	粗镗留下 0.25 mm 的余量	ISCAR TURNINGGOLDER MCLNR/L 20-4	CNMG-432-NR
9	09	精镗	ISCAR TURNING GOLDER MCLNR/L 20-4	CNMG-432-NF
10	10	切断	ISCAR TURNING GOLDER MCLNR/L 19-4	GFN 6

2.5　Setup Procedures for CNC Lathe

2.5　数控车床加工前的设置

The setup operation normally begins with the setup person securing the required OD and ID tools in the turret as specified in CNC tool and operations sheet. The part blank is loaded into the work-holding device (lathe chuck or lathe collet). It is very important that the proper length of stock extends beyond the work-holding device. If the overhand length is tools long, excessive wobbling will occur when the part is machined, and if the length is too short, the tool may collide with the work-holding device, or insufficient room may be left for the cut off tool to operate. The length of stock permitted to extend beyond the work-holding device is specified by the programmer in the setup sheet. Refer to Figure 2-7.

　　设置操作首先由操作工检查刀架上数控加工和刀具卡中指定的内外轮廓刀具，然后装夹零件坯料（车床卡盘或车床夹头）。装夹时，需要注意坯料的长度适当超出夹具，如果太长，在加工零件时会发生过度振动；如果太短，刀具可能会与卡盘碰撞，或者可能没有足够的长度切断工件。允许延伸到卡盘外的工件长度由程序员在零件夹具图中设定，如图 2-7 所示。

1. Tool Length Offsets

1. 刀具长度补偿

The setup person must measure and enter the value of the tool length offsets or geometry offset of each tool. The different tools used for machining operations in a program may vary in length and orientation in the turret. The system must be directed to compensate for these variations when executing programmed tool movement in the X and Z directions. It can only do so if it knows the initial distance between the tip of the tool and the program zero ($Z0$, $X0$). Tool length offsets are measured in relation to a reference tool. Each offset value is assigned a memory address number. Later, the controller will know the proper offset value for a tool when its memory address number is read during program execution. See Table 2-4.

操作工必须测量并输入每把刀具的长度补偿或几何补偿量的值。程序中用于加工操作的刀具在刀架中的长度和方向可以不同。当执行程序控制的刀具在 X 和 Z 方向上移动时,必须控制系统弥补这些补偿量。只有在知道刀具刀尖和编程原点($X0$, $Z0$)之间的初始距离时才能这样做。刀具长度补偿值是相对于参考刀具测量的。为每个补偿值分配一个存储器地址,之后,当程序执行期间读取其存储器地址编号时,控制器将会知道刀具的正确补偿值,见表2-4。

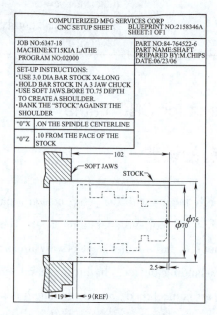

Figure 2-7 The CNC Setup Sheet for Setting Up the Part SHAFT

图2-7 数控零件夹具图

Table 2-4 Tool Offset Measurement and Point on Tool Programmed

Tool offset measurement(s)	Point on tool programmed
From part blank to tool edge(s)	Tool edge(imaginary tool nose radius) point located at the intersection of the horizontal and vertical tangents to the tool nose radius(TNR)
From part blank to center of tool nose	Center of tool nose point

表 2-4　刀具编程时刀具补偿测量

刀具补偿测量	刀具编程
从零件毛坯到刀刃	刀尖(假想的刀尖圆弧半径)点位于刀尖圆弧半径(TNR)的水平和垂直切线的交点处
从零件毛坯到刀尖圆角圆心	刀尖圆角圆心

（1）Manual Touch Off Method
（1）手动输入方式

The manual touch off method is the most common method of entering tool offset. The setup person switches the CNC controllers screen to POSITION-ABSOLUTE and sets the mode switch to JOG. A tool is selected as the reference tool, normally an OD finishing tool. It is indexed into position and then manually moved to the work. The tool is jogged in the X axis until its tip touches a specified diameter of the work. The setup person then zeroes out the X position (offset value) appearing on the screen. The tool is then jogged in the Z axis until its tip touches a specified face of the work. Again, the Z-position (offset value) appearing on the screen is zeroed out. Every tool is then indexed and manually touched off the diameter and face of the work. The corresponding X and Z position (offset) values are written down. After all values are recorded, the controller screen is switched to OFFSET. The reference tool is assigned zero offset values for the proper offset number. The offset values for each tool are also keyed in for the corresponding offset number of that tool.

手动输入方法是输入刀具补偿的最常用方法。操作工将数控机床的屏幕切换到POSITION-ABSOLUTE状态并将使用JOG模态。选择一把刀具作为参考刀具,这把刀通常是外圆精加工刀具。将该把刀具换到相应切削刀位,然后手动移动到工件。该刀具在 X 轴上进行微动,直到其尖端接触到的工件直径。然后,操作工将出现在屏幕上的 X 轴位置(偏置值)置零。然后将工具在 Z 轴上进行微动,直到其尖端接触到工件的端面。同样,出现在屏幕上的 Z 轴位置(偏置值)置零。然后将每把刀具调出并手动接触工件的直径和端面。写下相应的 X 轴和 Z 轴位置(偏置)值。记录完所有值后,屏幕切换到OFFSET界面。在相应位置给参考刀具设置零点补偿值。每把刀具的补偿值也键入该刀具的相应补偿值编号处。

This method is time consuming and requires an experienced person to complete properly. See Figure 2-8.

该方法较为耗时并且要求操作工具有丰富的经验,如图2-8所示。

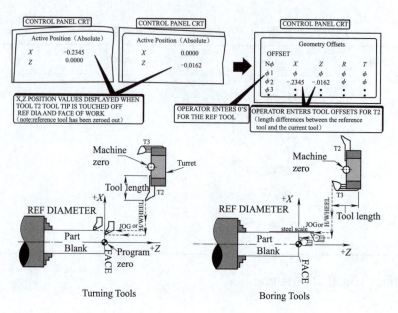

Figure 2-8　InputtingTool Offsets Via the Manual Touch off Method

图 2-8　通过手动方法输入刀具补偿

(2)"Q-setter" or Tool Setter Method

(2)"Q 设置"及刀具设定法

The "Q-setter" is an option available on newer CNC lathes and turning centers. The mechanism's arm swings out in front of the work-holding device. A touch sense probe is attached to the end of the arm. See Figure 2-9.

"Q 设置"是新型数控车床和车削中心的设置方法。机械臂在卡盘前摆动,接触感应器连接到机械臂的末端,如图 2-9 所示。

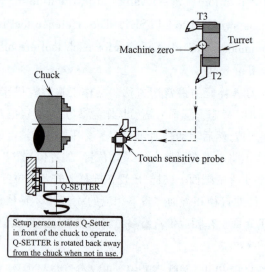

Figure 2-9　Inputting Tool Offsets Via Q-setter

图 2-9　通过 Q 设置输入刀具补偿

It offers several advantages over the manual method and makes the job of entering tool length offsets fast and accurate.

与手动方法相比，它具有多种优势，可以快速准确地输入刀具长度补偿值。

The control screen will automatically change to the GEOMETRY OFFSETS page when the setup person swings the "Q-setter" out into position. The number of the tool currently in position in the turret for probe touch off will be automatically highlighted. The control is set to JOG, and the tool is jogged in the X axis, then manually moved by the hand wheel. When its tip touches the center of X-axis sensor a beep will be emitted. The X-offset value will be automatically entered. The tool is then jogged in the Z-axis and touching the center of the Z-axis sensor a beep is emitted. The corresponding Z-offset value for the tool will be automatically entered.

当操作工将"Q 设置"移动到适当的位置时，控制屏幕将自动切换到 GEOMETRY OFFSETS 页面。当前在刀架中的切削位置的刀具编号将自动显示。切换到 JOG 模态，刀具在 X 轴上缓慢移动，然后由手轮手动移动。当其尖端接触 X 轴传感器的中心时，将发出蜂鸣声，将自动输入 X 轴补偿值。然后，刀具在 Z 轴上缓慢移动并接触 Z 轴传感器的中心，发出蜂鸣声，将自动输入刀具的相应 Z 轴补偿值。

(3) Wear Offsets

(3) 磨损补偿

The cutting surfaces of tools wear with use. As the tool wears, it leaves more material after machining a surface. For turning tools wear causes external dimension to grow. For boring bars, internal dimensions will get smaller. The operator opens the wear offset page at the CNC control panel and enters tool wear offsets for each tool to compensate for tool wear (Figure 2-10). This is done during production machining to meet tolerances.

刀具的切削刃会随着使用而逐渐磨损。当刀具磨损时，在加工表面之后会留下更多的多余材料。对于外圆车刀，磨损会导致外部尺寸增大。对于镗刀，内部尺寸将变小。操作人员打开数控机床控制面板上的磨损补偿页面，输入每个刀具的刀具磨损值，以补偿刀具磨损（图 2-10），输入磨损补偿值能够在生产加工期间满足公差要求。

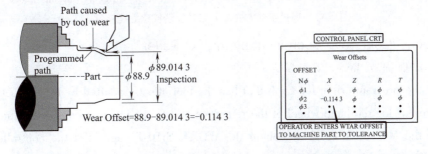

Figure 2-10　Inputting Tool Wear Offsets

图 2-10　输入刀具磨损补偿

2. Setting the Program Origin
2. 设置编程原点

(1) Tool Offsets Have been Inputted Manual Touch Off Method

(1) 手动方式输入刀具补偿值

When the manual touch off method has been used to input the tool offsets for a job, proceed as follows to set the program origin. Jog then hand wheel the reference tool to the position considered as the Z0 of the part. Open the MENU OFFSET screen to the to the WORK SHIFT page. Set the controller to MDI mode. Key in G50 Z0 and press ENTER. The POSITION-ABSOLUTE screen will display the Z position of tool as Z0. Now measure the outside diameter of the stock D. Jog then hand wheel for the reference tool to the outside diameter and touch off the surface. Set the controller to MDI mode again, key in G50 X_d, and press ENTER. The POSITION-ABSOLUTE screen will now display X_d Z0. See Figure 2-11.

当手动方法被用于输入刀具补偿值时,请按以下步骤设置编程原点。选择 JOG 模式,然后使用手轮将参考刀具运动到零件 Z0 的位置,打开 MENU OFFSET 界面,进入 WORK SHIFT 页面,将控制器设置为 MDI 模式,键入 G50 Z0 并按 ENTER 键,界面上将刀具的 Z 绝对位置显示为 Z0。然后测量工件的外径 D,用手轮将刀具转到外径并接触工件表面,再次将控制器设置为 MDI 模式,键入 G50 X_d,然后按 ENTER 键。POSITION-ABSOLUTE 界面上 X 和 Z 轴位置将显示 X_d Z0,如图 2-11 所示。

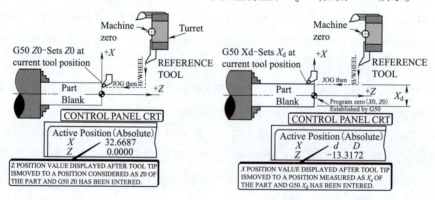

Figure 2-11　Establishing the Program Zero with Tool Offsets Input by Manual Touch off

图 2-11　通过手动方式输入刀具补偿值建立程序零点

(2) Tool Offsets Have Been Input by the "Q-Setter"

(2) 通过"Q-setter"输入刀具补偿值

To set the program origin for a job that has the tool offsets inputted by the "Q-setter", proceed as follows. Jog the hand wheel of the reference tool to the position considered the Z0 of the part. Open the MENU OFFSET screen to the WORK SHIFT page. Set the controller to MDI mode. Key in Z0 and press ENTER. The POSITION-ABSOLUTE screen will display the Z position of the tool as Z0, X is set by default since the center of rotation is considered X0 regardless of the part diameter. See Figure 2-12.

已用 "Q 设置" 输入刀具补偿时,设置程序零点,请执行以下操作。将参考刀具的手轮点动到零件 Z0 的位置,在屏幕上打开 MENU OFFSET 的 WORK SHIFT 页面,改为 MDI 模式,键入 Z0 并按 ENTER 键。POSITION-ABSOLUTE 界面将显示刀具的 Z 轴坐标位置设为 Z0,X 轴默认设置,因为无论零件直径多大,旋转中心都被视为 X0,如图 2-12 所示。

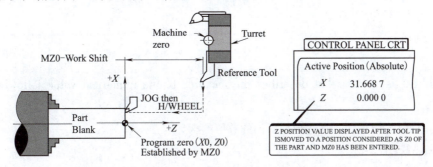

Figure 2-12　Establishing the Program Zero with Tool Offsets Input from the Q-setter

图 2-12　通过 Q 设置方法输入刀具补偿值建立程序零点

2.6　Programming Language Format

2.6　程序格式

A program format is a system of arranging information so that it is suitable for input to a CNC controller. The current standard for CNC programming is based on ISO 6983. Several different types of format exist. The ANSI/ERA 274-D-1980: Interchangeable Variable Block Data Format, also known as Word Address format will be used in this book. It was originally developed for use with NC tapes and has been retained for CNC programming. Programming of CNC equipment involves the entry of word address code for precisely controlling all machine movements.

程序格式是适合输入数控机床控制器的一种有序信息。当前数控编程采用 ISO 6983 标准,并有几种不同类型的格式。ANSI / ERA 274-D-1980:字、地址可变程序段格式,也称为字地址格式,将在本书中使用。它最初开发用于 NC 磁带,现仍用于数控编程。数控设备的编程包括输入字地址码,以精确控制所有机床运动。

1. Programming Language Terminology

1. 编程术语

The following terminology is important when using the word address format.

使用字地址格式时,以下术语很重要。

(1) Programming Character

(1) 程序字符

A programming character is an alphanumeric character or punctuation mark.

程序字由字母、数字、字符或标点符号构成。

■ **EXAMPLE 2-1**

The following are programming character: N G.

■ **例 2-1**

以下为程序字符: N G。

(2) Addresses

(2) 地址符

An address is a letter that describes the meaning of the numerical value following the address.

地址符是一个字母,用于描述地址后面的数值的含义。

■ **EXAMPLE 2-2**

Identify the address and the number in the codes G00 and X-3.75:

■ **例 2-2**

识别 G00 和 X-3.75 中的地址符和数值:

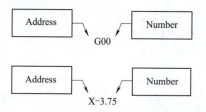

It is important to note that a minus (-) sign may be inserted between the address and the numeric value. Positive values do not need a plus sign.

需要注意的是在地址符和数值之间插入减号(-),而正值不需要特别标明。

(3) Words

(3) 字

Characters are used to form words. Program words are composed of two main parts: an address followed by a number. Words are used to describe such important information as machine motions and dimensions in programs.

字符构成字。程序字由两个主要部分组成:地址后跟数字。字用于描述诸如机床运动和程序中的尺寸之类的重要信息。

(4) Blocks

(4) 程序段

A block is a complete line of information to the CNC machine. It is composed of one word or an arrangement of words. Blocks may vary in length, thus, the programmer need only include in a block those words required to execute e a particular machine function.

一行程序段是数控机床的一条完整信息,它由一个字或一段有序排列的字组成。程序段的长度会变化,因此,程序员只需要在程序段中输入特定机床功能所需的那些字。

EXAMPLE 2-3

Point out the components of the block N0020 X-2.5 Y3.75 S1000:

例 2-3

指出程序段 N0020 X-2.5 Y3.75 S1000 的组成：

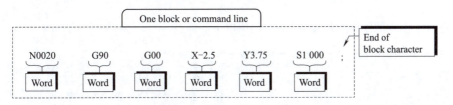

程序的输入方法3D视频

Each block is separated from the next by an end-of-block（;）code.

每个程序段通过程序段结束符(;)与下一个程序段分开。

NOTE：The end-of-block character is automatically generated when the programmer enters a carriage return at the computer. The same holds true when the end-of-block key is depressed at the machine control unit during manual entry. Therefore, this character will not appear in the regular program listings in this text.

注意：当程序员在计算机上输入回车时，会自动生成程序段结束字符。当在手动输入期间在机床控制单元处按下程序段结束键时也是如此。因此，此字符不会出现在文本中常规程序清单列表中。

(5) Programs

(5) 程序

A program is a sequence of blocks that describe in detail the motions a CNC machine is to execute in order to manufacture a part. The MCU executes a program block by block. The order in which the blocks appear is the order in which they are processed.

程序编辑操作

程序是一系列程序段，详细描述了数控机床为了加工零件而执行的运动。MCU 逐段执行程序，程序段排列的顺序就是它们的加工顺序。

EXAMPLE 2-4

Illustrate the order in which the MCU executes the following program.

例 2-4

说明 MCU 执行以下程序的顺序。

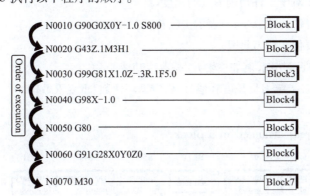

图形模拟操作

2. Arrangement of Addresses in a Block

2. 程序段中的地址符排列顺序

The order in which addresses appear in a block can vary. The following sequence, however, is normally used:

地址符在程序段中出现的顺序可以变化,通常使用以下顺序:

①→②→③→④→⑤→⑥→⑦→⑧→⑨→⑩→⑪→⑫→⑬→⑭
N.. G.. X.. Y.. Z.. I.. J.. K.. U.. (V.. W.. A.. B.. C)
⑮→⑯→⑰→⑱→⑲→⑳→㉑→㉒
P.. Q.. R.. F.. S.. T.. M.. H

And each address means:

(1) N Sequence number, indicates the sequence number of the block.

(2) G Preparatory function, specifies the mode of operation in which a command is to be executed.

(3) X, Y, Z (I, J, K; U, V, W; A, B, C; P, Q, R) Dimension words, designate the amounts of axis movements.

(4) F Feed rate, designates the relative speed of the cutting tools with respect to the work.

(5) S Spindle function, designates the spindle speed in revolutions per minute (r/min).

(6) T Tool function, designates the number of the tools to be used.

(7) M Miscellaneous function, designates a machine function such as spindle on/off or coolant on/off.

(8) H, D Auxiliary input function, specifies tool length offset number, number of, repetitions of a fixed cycle, and so on.

每个地址符的含义是:

(1) N 程序段号,表示程序段的序号。

(2) G 准备功能字,指定执行命令的操作模式。

(3) X、Y、Z (I、J、K; U、V、W; A、B、C; P、Q、R) 坐标值,表示轴的移动量。

(4) F 进给功能字,表示相对于工件切削刀具的速度。

(5) S 主轴功能字,以每分钟转数(r/min)表示主轴转速。

(6) T 刀具功能字,指定要使用的刀具号。

(7) M 辅助功能字,指定机床功能,例如主轴开/关或冷却液开/关。

(8) H、D 辅助输入功能字,指定刀具长度补偿号、固定循环次数、重复次数等。

■ **EXAMPLE 2-5**

Give an example of address arranged in a block.

■ 例 2-5

给出一条典型的程序段示例。

2.7　Important Preparatory Functions（G Codes）for Lathe

2.7　数控车床的重要准备功能代码（G 代码）

The following modal and nonmodal G codes are important when programming lathe operations. See Table 2-5.

下面是车床加工编程中重要的模态和非模态 G 代码，见表2-5。

Table 2-5　The Modal and Nonmodal G codes

G code	Mode	Specification
G0	Modal	Rapid positioning mode. The tool is moved to the programmed X and Z position at maximum feed rate
G1	Modal	Linear interpolation mode. The tool is moved in straight-line path at programmed feed rate
G2	Modal	Circular interpolation clockwise（CW）
G3	Modal	Circular interpolation counterclockwise（CCW）
G4	Nonmodal	Programmed dwell
G20	Modal	Inch mode for all units. This code is entered at the start of the CNC program in a separate block. It must appear before a G50 block
G21	Modal	Metric mode for all units. This code is entered at the start of the CNC program in a separate block. It must appear before a G50 block
G28	Nonmodal	Return to tool reference point
G50	Modal	Maximum spindle speed setting for G96 mode. Or Set machining origin at the current tool position
G75	Nonmodal	Grooving in the X-axis
G96	Modal	Constant surface speed control（m/min）
G97	Modal	Cancel G96
G98	Modal	The feed rate is in units of millimeters per minute
G99	Modal	The feed rate is in units of millimeters per revolution

表2-5 模态和非模态 G 代码

G 代码	模式	含义
G0	模态	快速定位指令。刀具以最大进给量移动到编程的 X 轴和 Z 轴位置
G1	模态	线性插值指令。刀具以编程的进给量沿直线路径移动
G2	模态	顺时针圆弧插补(CW)指令
G3	模态	逆时针圆弧插补(CCW)指令
G4	非模态	暂停指令
G20	模态	所有单位的英制模式。该代码在数控程序开始时在单独的程序段中输入。它必须出现在 G50 之前
G21	模态	所有单位的公制模式。该代码在数控程序开始时在单独的程序段中输入。它必须出现在 G50 之前
G28	非模态	返回刀具参考点
G50	模态	G96 模式的最大主轴转速设置,或者在当前刀具位置设置机床原点
G75	非模态	在 X 轴上开槽
G96	模态	恒线速度控制,单位为 m/min
G97	模态	取消 G96
G98	模态	进给量以 mm/min 为单位
G99	模态	进给量以 mm/r 为单位

2.8　Important Miscellaneous Functions（M Codes）for Lathe

2.8　数控车床重要的辅助功能代码（M 代码）

视频
M指令的使用

The following miscellaneous functions are often used to initiate machine functions not related to dimensional or axis movements, See Table 2-6.

以下辅助功能通常用于启动与尺寸或轴运动无关的机床功能,见表2-6。

Table 2-6　The Miscellaneous Functions

M Code	Mode	Specification
M0	B	Causes a program stop
M3	A	Turns spindle on clockwise（CW）
M4	A	Turns spindle on counterclockwise（CCW）
M5	B	Turns spindle off
M8	A	Turns external coolant on
M9	B	Turns coolant off
M30	B	Directs the system to end program processing, and reset the memory unit. This code must be the last statement in a program

Continued

M Code	Mode	Specification
M41	A	Shifts spindle into low-gear range
M42	A	Shifts spindle into intermediate-gear range
M43	A	Shifts spindle into high-gear range
M68	A	Clamps the chuck
M69	B	Opens the chuck

表 2-6　辅助功能字

M 代码	类型	规范
M0	B	程序停止
M3	A	顺时针转动主轴（CW）
M4	A	逆时针转动主轴（CCW）
M5	B	关闭主轴
M8	A	打开外部冷却液
M9	B	关闭外部冷却液
M30	B	系统结束程序，重置存储器单元。此代码必须是程序中的最后一个语句
M41	A	将主轴切换到低速挡
M42	A	将主轴切换到中速挡
M43	A	将主轴切换到高速挡
M68	A	夹紧卡爪
M69	B	松开卡爪

2.9　Setting the Machining Origin

2.9　建立机床原点

The machining origin or tool change position is established once at the beginning of the word address program before any block calling for a tool change. The setup person rotates the longest OD tool into position in the turret. The tool is then moved to 25 mm of the reference diameter by using jog or the hand wheel. The X value a displayed on the POSITION-ABSOLUTE screen of the control is written down. The longest ID tool is then rotated into position in the turret. Again, the tool is moved to 25 mm of the face of the stock by jogging or hand wheeling. The Z value b displayed on the POSITION-ABSOLUTE screen of the control is also written down. These are the X, Z values used to establish the machining origin or tool change position. This is accomplished by entering the block G50 Xa Zb once in the word address program before any block calling for a tool change. See Figure 2-13.

在任何要求换刀的程序段之前,字地址程序开始时,建立一次机床原点或换刀位置。操作工将最长的外圆切削刀具旋转到刀架中加工位置,然后使用寸动模式或手轮将刀具移动到距离参考直径的 25 mm 的位置,记下 POSITION-ABSOLUTE 画面上显示的 X 轴的数值 a。然后将最长的内轮廓切削刀具旋转到刀架中的加工位置,接着通过寸动模式或手轮将刀具移动到毛坯端面 25 mm 处,在 POSITION-ABSOLUTE 画面上显示的 Z 轴的值 b 也记录下来。这两个数值是用于确定机床原点或换刀位置的 X 轴、Z 轴的值。在程序中,换刀程序之前通过输入 G50 Xa Zb 来确定机床原点,如图 2-13 所示。

All subsequent tool movements will then be computed by the controller relative to the machining origin established.

然后,控制器将相对于建立的机床原点计算所有后续刀具的坐标。

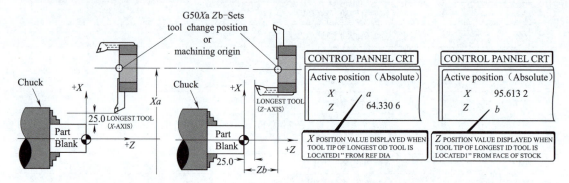

Figure 2-13　Specifying the Location of Machine Origin from Program Zero

图 2-13　从编程原点指定机床原点

■ **EXAMPLE 2-6**

Enter a program block for creating a machining origin at the location shown (Figure 2-14) below.

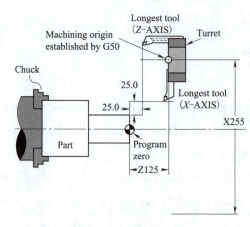

Figure 2-14　Case 2-6

图 2-14　例 2-6

The Program Block for Creating a Machining Origin, See Table 2-7.

Table 2-7 The Program Block for Creating a Machining Origin

	Word address command	Meaning
G50 is coded before first block in the program that calls for a tool change	N0020 G50 X225. Z125.	Set machining origin (tool change position) at X255. Z125
	N0030 T0101	Return to tool change position. Index to tool 1. Cancel values in offset file 1

■ 例 2-6

编写一段程序用于创建图 2-14 中机床原点。

创建加工原点的程序段见表 2-7。

表 2-7 创建加工原点的程序段

	指令	含义
在换刀之前完成 G50 指令	N0020 G50 X225. Z125.	在 X225. Z125. 的位置设置机床原点（换刀位置）
	N0030 T0101	返回换刀位置。换 1 号刀，取消刀具补偿文件 1 号位的值

2.10 Feed Rate（F Code）

2.10 进给功能（F 代码）

The numerical value following the address F specifies the feed rate. When programmed with the preparatory function G99, it indicates the tool mm/r.

F 后面的数值指定进给量。使用预备功能 G99 进行编程时，表示刀具进给量为 mm/r。

 General Syntax

G99-Fn

G99 The feed rate is in millimeters per revolution.

Fn The values of n specifies the tool feed in mm/r.

 常用格式

G99-Fn

G99 指定刀具进给量的单位。

Fn n 的值指定刀具以 mm/r 为单位的刀具进给量。

2.11 Spindle Speed (S Code)

2.11 主轴功能(S代码)

The spindle speed is specified by the S code. When programmed with the preparatory function G97, it indicates the spindle r/min.

主轴转速由S代码指定。使用预备功能G97进行编程时,表示主轴转速为r/min。

General Syntax

G97-Sn

G97 Cancels any previous constant surface speed control on the spindle.

Sn The values of n specifies the spindle speed in r/min.

常用格式

G97-Sn

G97 取消主轴上任何先前的恒线速度控制。

Sn n的值指定以r/min为单位的主轴转速。

■ EXAMPLE 2-7

Program a spindle speed of 1600 r/min.

N0050 G97S 1600

■ 例2-7

编写一段编程,主轴转速为1600 r/min。

N0050 G97 S 1600

2.11.1 Spindle Speed With Constant Surface Speed Control

2.11.1 主轴恒定线速度控制

Recall that the cutting speed is given by the formula:

$$\text{Cutting speed} = (\pi \times D \times \text{r/min})/1\,000$$

Where r/min is the spindle speed and D is the work diameter.

切削速度由下式给出:

$$\text{切削速度} = (\pi \times D \times \text{主轴转速})/1\,000$$

其中r/min是主轴转速,D是切削直径。

As the tool removes material from the outside, the work diameter will decrease. Thus, the cutting speed will also decrease. The opposite will be true for inside operations with the cutting speed increasing as the machining progresses. Tool suppliers recommend specific cutting speeds to be maintained in order for tools to operate at optimum performance and produce a required surface finish. The control can be directed to adjust the r/min of the spindle such that a constant surface speed results as the part diameter changes. A G96 code can be programmed to ensure constant surface speed control. An S word entered with this code no longer indicates spindle speed but tool cutting speed.

当刀具从外部切削材料时,工件直径将减小。因此,切削速度也将降低。对于切削内部材料,情况正好相反,切削速度随着加工的进行而增加。刀具供应商提供特定的切削速度,保证刀具以最佳性能运行并产生所需的表面粗糙度。可以调节主轴的转速,使得随着工件直径变化而产生恒定的线速度。G96 指令可以确保恒定的线速度控制。使用此代码输入的 S 不再表示主轴转速,而是表示刀具切削速度。

General Syntax

G96 Sn

G96 Specifies constant surface speed to be in effect.

Sn n value represents the tool cutting speed in meter per miniate. This value is to be maintained as the part diameter changes.

常用格式

G96 Sn

G96 指定恒线速度有效。

Sn n 值表示以 m/min 为单位的刀具切削速度。零件直径发生变化时,该值保持不变。

■ EXAMPLE 2-8

Write a program block directing the controller to adjust the spindle RPM so that a constant surface speed of 600 m/min is held constant during machining.

N0060 G96 S600

■ 例 2-8

编写一个程序段,调整主轴转速,加工过程中保持恒线速度为 600 m/min。

N0060 G96 S600

The controller is now instructed to adjust the spindle rpm upward if the tool moves to a smaller diameter and downward if the tool moves to a larger diameter.

控制器将随着直径减小提高主轴转速,直径变大减小主轴转速。

2.11.2 Spindle Speed With Clamp Speed and Constant Surface Speed Controls

2.11.2 恒定线速度控制

The G96 command the controller to increase or decrease the spindle rpm. There may be cases where a specific rpm level is not to be reached, such as if there are chucking requirements or the work becomes unstable at certain rpm. A G50 code will assign an upper limit on the rpm value the controller is not to exceed. when S is coded with G50 it indicates spindle r/min.

G96 指令控制控制器增加或减少主轴转速。可能存在不能达到特定转速水平的情况,例如,如果存在夹紧要求或者工作在某些转速下变得不稳定。G50 代码将指定控制器控制主轴转速不超过特定的上限值。当 S 用 G50 编码时，它表示主轴转速 r/min。

General Syntax

G50 Sn

G50 Specifies that the controller is not exceed a programmed RPM value.

Sn The value of n indicates the maximum spindle RPM not to be exceed.

常用格式

G50 Sn

G50 指定控制器不能超过指定的主轴转速值。

Sn n 的值表示最大主轴转速不得超过的速度。

EXAMPLE 2-9

For the facing to center operation shown in Figure 2-15, the spindle is to start at 700 r/min and is not to exceed 1 500 r/min. Write blocks to assign constant surface speed control under these conditions.

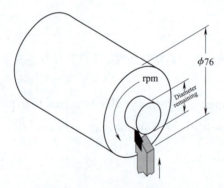

Figure 2-15　Case 2-9

图 2-15　例 2-9

$$\text{Cutting speed} = (\pi \times D \times \text{rpm})/1\,000$$

$$\text{Cutting speed} = (\pi \times 76 \times 700)/1\,000$$

$$\text{Cutting speed} = 167 \text{ m/min}$$

$$切削速度 = (\pi \times 切削直径 \times 主轴转速)/1\,000$$

$$切削速度 = (\pi \times 切削速度 \times 700)/1\,000$$

$$切削速度 = 167 \text{ m/min}$$

N0020 G50 S1500

N0100 G96 S167

例 2-9

对于图 2-15 中所示的加工，主轴应以 700 r/min 的转速启动，且不超过 1 500 r/min。在此条件下按恒定线速度控制编写程序。

N0020 G50 S1500

N0100 G96 S167

The following table (Table 2-8) lists the spindle RPM values that the controller will assign as the work diameter decreases.

表2-8列出了当工件直径减小时控制器将分配的主轴转速值。

Table 2-8 The Diameter and Spindle Speed

Diameter	r/min
76	700
60	886
35.4	1500
30	1500

表2-8 直径及主轴转速

直径	主轴转速(r/min)
76	700
60	886
35.4	1 500
30	1 500

The controller will continue to increase the spindle RPM as the diameter decreases to 60mm. At diameters less than 35.4mm, the maximum spindle RPM will be reached and the controller will assign this value as the tool cuts to the spindle center line. Thus, constant surface speed will be maintained as the tool cuts from 76mm to 35.4mm and will decrease thereafter.

当主轴直径减小到60 mm时,控制器将继续增加主轴转速。当直径小于35.4 mm时,将达到最大主轴转速,控制器将一直使用最大转速,直到切削直径为0。因此,当刀具从76 mm切削到35.4 mm时,将保持恒定的线速度并且此后将减小。

2.12 Automatic Tool Changing

2.12 自动换刀功能

A general word form is used to program tool changes is：

用于换刀的常用表达指令。

 **General Syntax**

Tab

T Specifies a tool change.

a Is a two-digit number indicating the turret station where the new tool is located b Is a two-digit number indicating the memory address where the new tool's offset values are stored.

常用格式

Tab

T 为更换刀具功能字。

a 是一个两位数字，表示新刀具所在刀架中的位置。

b 是一个两位数字，表示存储刀具补偿值的存储器地址。

Prior to executing a tool change, the programmer should enter a command to return to the tool change position and cancel the all offsets of the old tool used.

在执行换刀操作之前，程序员应输入命令返回换刀位置并取消所用旧刀具的刀具补偿值。

■ EXAMPLE 2-10

Code blocks directing the controller to change from tool 1 to tool 4. See Figure 2-16.

■ 例 2-10

程序段控制控制器将刀具 1 换为刀具 4，如图 2-16 所示。

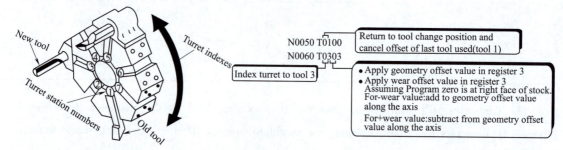

Figure 2-16　Case 2-10

图 2-16　例 2-10

The control will properly move tool 4 after the tool change because it will know the corresponding tool offsets.

换刀后，由于控制器知道刀具 4 的刀具补偿值，因此能将刀具 4 移动到正确位置。

2.13　Tool Edge Programming

2.13　刀尖编程方式

Tool edge programming is in effect when offsets are measured to the edge of the tool. With tool edge programming, the part geometry may be input directly. Because tool edge programming ignores the tool nose radius of cutting tools, it can only be applied to part geometries consisting of horizontal or vertical lines. Errors will result if arcs or tapered lines are programmed without applying additional compensation. See Figures 2-17 and 2-18.

当测量到刀具刀尖的补偿量时，刀具刀尖编程有效。通过刀具刀尖编程，可以直接输入零件的几何形状。由于刀具刀尖编程忽略了刀具的刀尖圆弧半径，因此它只能应用于由水平或垂直线组成的零件几何形状。如果编程圆弧或锥形线而不应用额外补偿值，则会导致错误，如图 2-17 和图 2-18 所示。

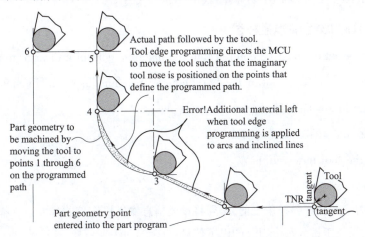

Figure 2-17　Programming the Part Geometry Directly with Tool Edge Programming

图 2-17　使用刀具刀尖编程直接编程零件几何图形

The inherent problems with tool edge programming can be corrected if tool nose radius (TNR) compensation is applied.

如果应用刀尖圆弧半径（TNR）补偿，则可以纠正刀具刀尖编程的问题。

2.14　Tool Nose Radius Compensation Programming

2.14　刀尖圆弧半径补偿编程方式

For CNC lathes and turning centers, this feature is called tool nose radius (TNR) compensation. See Figure 2-18.

对于数控车床和车削中心，此功能称为刀尖圆弧半径（TNR）补偿，如图 2-18 所示。

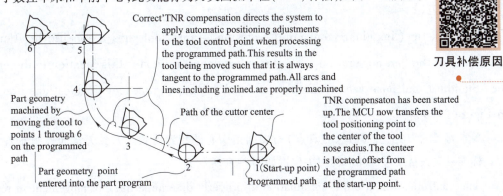

Figure 2-18　Programming the Part Geometry Directly with TNR Programming

图 2-18　使用 TNR 编程直接编程零件几何图形

2.14.1　Setting Up Tool Nose Radius Compensation

2.14.1　设置刀具刀尖圆弧半径补偿

Important information must be entered into the machine control unit prior to using TNR compensation. During setup, the control is instructed to open an offset file in memory. The setup person keys in the following information for each tool:

(1) X and Z tool offsets to the tool edge.

(2) Size of the tool nose radius.

(3) Tool nose vector.

在使用刀具刀尖圆弧半径补偿之前，必须将重要信息输入机床控制单元。在操作时，控制器在内存中打开文件，操作工键入每个刀具的以下信息：

(1) 刀具 X 和 Z 向的刀具刀尖补偿值。

(2) 刀尖圆弧半径的值。

(3) 刀尖矢量。

1. Tool Nose Radius

1. 刀尖圆弧半径

The size of the tool nose radius is readily available from tool supplier catalogs or the insert package. Standard radii are 0.396 875 mm or 0.396 24 mm, 0.793 75 mm or 0.792 48 mm, and 1.190 625 mm or 1.191 26 mm.

刀尖圆弧半径的大小可从刀具供应商目录或刀片包装盒中获得。标准半径为 0.396 875 mm 或 0.396 24 mm、0.793 75 mm 或 0.792 48 mm、1.190 625mm 或 1.191 26 mm。

2. Tool Nose Vector

2. 刀尖矢量

The tool tip (imaginary tool point) of single-point tools has a specific location from the center of the tool nose radius. The tool nose vector indicates this location to the controller. Standard tool nose vector numbers and the corresponding tool tip locations are shown in Figure 2-19.

单点刀具的刀尖位于（假想刀具点）刀尖圆角中心的特定位置。刀尖矢量向控制器表明该位置。标准刀尖矢量编号和相应的刀尖位置如图 2-19 所示。

The controller will use this information to properly determine the tool movement in response to a compensation command.

控制器将使用该信息响应补偿命令来正确确定刀具运动。

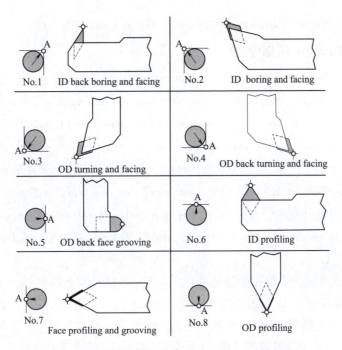

Figure 2-19 Tools Nose Vectors and Tool Tip Locations

图 2-19 刀具刀尖矢量和刀尖位置

2.14.2 Some Restrictions With Tool Nose Radius Compensation

2.14.2 刀尖圆弧半径补偿的注意事项

Some restrictions directly apply to the Fanuc family of controllers; they are also valid for other types of controllers.

(1) A G code activating TNR compensation is entered as a separate block in the program. It must be commanded before the tool starts to cut.

(2) The first or second block following a TNR compensation code must contain an X, Z linear motion command. The motion command signals the controller to initiate TNR compensation or cancel compensation. Failure to do this will result in over or undercutting.

(3) Motion in the G40, G41, or G42 block must be greater than twice the value of the tool nose radius. Following this rule will ensure that overcutting or undercutting will not result.

以下注意事项适用于 Fanuc 系列控制器和其他类型的控制器。

(1) 激活刀尖圆弧半径补偿的 G 代码在程序中的单独程序段输入。必须在刀具开始切削之前输入指令。

(2) 刀尖圆弧半径补偿指令输入后的第一个或第二个程序段必须包含 X, Z 轴直线运动指令。运动过程中指令向控制器发出信号以启动刀尖圆弧半径补偿或取消补偿。如果不设置,将导致过切或欠切。

(3) G40、G41 或 G42 程序段中的移动量必须大于刀尖圆弧半径值的两倍,以确保不过切或欠切。

2.14.3 Tool Nose Radius Compensation Commands
2.14.3 刀尖圆弧半径补偿指令

A G41 or G42 word indicates TNR compensation is to start up on the next linear move. When G41 or G42 is active, the MCU transfers the tool positioning point from the imaginary point A as shown in Figures 2-20 and 2-21 to the center of the tool nose radius. The controller will signal for the tool to be offset at the start-up point by the radius value stored in the MCU's offset file. The tool nose vector number for the particular tool used is also stored in the offset file. This number enables the MCU to determine the location of the tool edge. Automatic adjustments will than be made such that the tool nose is always positioned tangent to the programmed path. When TNR compensation is canceled with a G40 word, the MCU shifts the tool positioning point back to the imaginary point A.

G41 或 G42 指令指定刀尖圆弧半径补偿表是在下一次线性移动时启动。当 G41 或 G42 激活时，MCU 将刀位点从假想点 A（见图 2-20 和图 2-21）传送到刀尖圆角的中心。控制器按照存储在 MCU 补偿文件中半径补偿值在启动点处移动刀具，使用的刀具刀尖矢量编号也存储在补偿文件中，此编号使 MCU 能够确定刀具刀刃的位置，自动调整刃口位置，使得刀尖总是与编程路径相切。当用 G40 取消刀尖圆弧半径补偿时，MCU 将刀位点移回假想点 A。

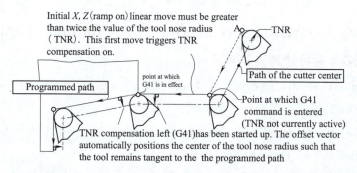

Figure 2-20　Ramp on Linear Move to Initiate Tool Nose Radius Compensation Left.

图 2-20　使用线性移动以启动刀尖圆弧半径左补偿

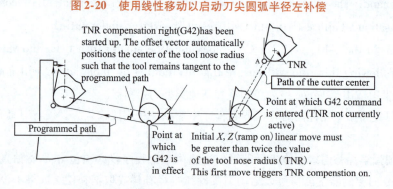

Figure 2-21　Ramp on Linear Move to Initiate Tool Nose Radius Compensation Right.

图 2-21　使用线性移动以启动刀尖圆弧半径右补偿

1. Tool Nose Radius Compendation Left

1. 刀尖圆弧半径左补偿

 **General Syntax**

G41 Xn Zn Tab

G41 Direct the controller to offset the tool to the left of the tool motion. The offset will occur on the next linear X,Z-axis move. See Figure 2-22.

Tab Specifies the tool change.

a Is a two-digit number indicating the turret station number where the new tool is located.

b Is a two-digit number indicating the memory address where the new tool is located.

常用格式

G41 Xn Zn Tab

G41 控制控制器将刀具偏移到刀具运动的左侧。刀具补偿将在下一个线性 X、Z 轴移动时启动，如图 2-22 所示。

Tab 指换刀指令。

a 是一个两位数字，表示新刀具所在的刀架号。

b 是一个两位数字，表示新刀具所在的内存地址。

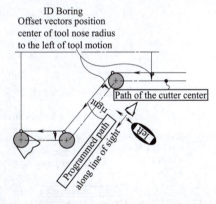

Figure 2-22　Tool Nose Radius Compendation Left

图 2-22　刀尖圆弧半径左补偿

2. Tool Nose Radius Compendation Right

2. 刀尖圆弧半径右补偿

G42 Xn Zn Tab

G42 Direct the controller to offset the tool to the right of the tool motion. The offset will occur on the next linear X,Z-axis move. See Figure 2-23.

G42 控制控制器将刀具偏移到刀具运动的右侧。刀具补偿将在下一个线性 X、Z 轴移动时发生，如图 2-23 所示。

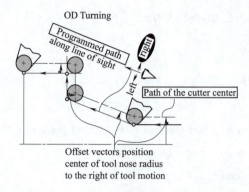

Figure 2-23　Tool Nose Radius Compendation Right

图 2-23　刀尖圆弧半径右补偿

3. Tool Nose Radius Compensation Cancel

3. 取消刀尖圆弧半径补偿

General Syntax

G40Xn Zn

G40 Cancels TNR compensation (G41 or G42). The controller will change the tool to an uncompensated position on the next linear X, Z-axis move. The linear move can be rapid (G0) or at feed rate (G1), See Figure 2-24.

常用格式

G40Xn Zn

G40 取消刀尖圆弧半径补偿(G41 或 G42)。控制器将在下一个线性 X、Z 轴移动时将刀具改为没有补偿的位置。线性移动可以是快速定位(G0)或直线插补(G1)运动,如图 2-24 所示。

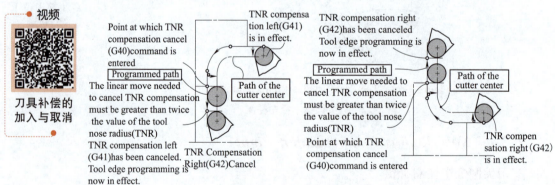

Figure 2-24　Tool Nose Radius Compensation Cancel

图 2-24　取消刀尖圆弧半径补偿

NOTES:

(1) Some controllers use single-digit numbers for a and b.

(2) More than one b code can be programmed with a tool. This allows for the use of the

same tool for executing rough as well as finish cuts or to compensate for tool wear.

(3) G2/G3 motion cannot be used to initiate a G41, G42, or G40 word in a program.

(4) G41/G42 is modal. This means they remain in effect for all subsequent tool motions until canceled by a G40 word. G40 is also modal and remains in effect until canceled by a G41/G42 word.

(5) The initial state of the control at machine start-up is G40.

注意事项:

(1) 有些控制器 a 和 b 是一位数。

(2) 一把刀具可以使用多个 b 代码编程。这样使用相同的工具可以来执行粗加工、精加工切削或补偿刀具磨损。

(3) G2/G3 动作不能用于在程序中启动 G41、G42 或 G40。

(4) G41/G42 是模态的。这意味着它们对所有后续刀具运动保持有效,直到被 G40 指令取消。G40 也是模态的并且在被 G41／G42 取消之前一直有效。

(5) 机床启动时控制的初始状态为 G40。

2.15　G Code

2.15　G 指令

2.15.1　Linear Interpolation Commands

2.15.1　直线插补指令

For CNC lathes, linear interpolation involves moving the tool along a straight line programmed at the specified feed rate. Linear interpolation (Figure 2-25) is used to execute such operations as facing, turning, taper turning, and taper boring.

对于数控车床,直线插补是按指定的进给量直线移动刀具。直线插补(见图 2-25)用于执行车削端面、车削外圆、锥度车削和锥度镗孔等操作。

In the discussion to follow it is assumed that tool nose radius compensation, TNR, is in effect.
在以下的论述中,假设刀尖圆弧半径补偿有效。

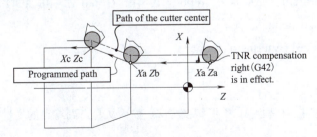

视频

G01直线
插补指令
2D轨迹

Figure 2-25　The Linear Interpolation Mode

图 2-25　直线插补指令

General Syntax

G1 X1 Z1 Fn

Xa Za

Xa Zb

Xc Zc

⋮

Xn Zn

G1 Specifies the linear interpolation mode.

Xa Za, Xb Zb, Xc Zc , ⋯ ,Xn Zn The values a, b,⋯ n specify the absolute coordinates of the programmed path when TNR is in effect.

Fn The value of n specifies the feed rate of the tool. If not programmed, the system will use the last feed rate programmed. If not specified at the beginning of the program, the system will issue an alarm.

常用格式

G1 X1 Z1 Fn

Xa Za

Xa Zb

Xc Zc

⋮

Xn Zn

G1 指直线插补指令。

Xa Za, Xb Zb, Xc Zc ⋯Xn Zn a, b,⋯ n 的数值指刀尖圆弧半径补偿生效时编程路径的绝对坐标。

Fn n 指刀具进给量。如果未编程指定，系统将使用程序中前面最近一次指定的进给量。如果未在程序开头指定，系统将发出警报。

NOTES：

(1) X-coordinate values are in terms of diameter.

(2) If incremential coordinates are used, replace X with U and Z with W. Input directed distance from start point to end point.

(3) U-coordinate values are in terms of radius.

注意事项：

(1) X 坐标值以直径表示。

(2) 如果使用增量坐标，则将 U 替换 X，将 W 替换为 Z。输入从起点到终点的增量值。

(3) U 坐标值以半径表示。

■ **EXAMPLE 2-11**

Write linear interpolation blocks to move the tool from position a to position b in each matching case shown in Figure 2-26.

■ **例 2-11**

使用直线插补指令,将刀具从位置 a 移动到位置 b,如图 2-26 所示。

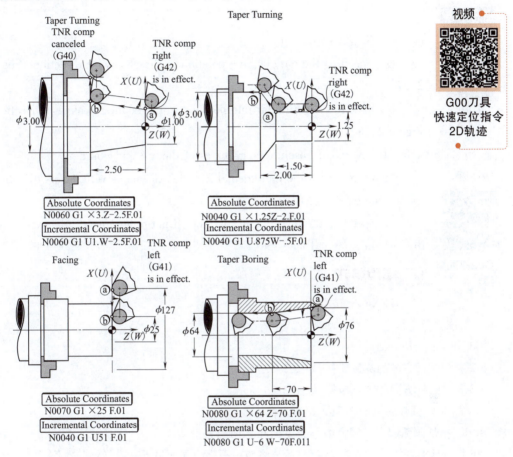

Figure 2-26 Case 2-11

图 2-26 例 2-11

2.15.2 Circular Interpolation Commands

2.15.2 圆弧插补指令

Circular interpolation for lathe operations involves cutting a circular arc in either a clockwise or counterclockwise direction at the programmed feed rate.

用于车床操作的圆弧插补是以指定的进给量沿顺时针或逆时针方向切削圆弧。

Assume, again, that tool nose radius compensation TNR is in effect.

再次假设刀尖圆弧半径补偿有效。

1. Center-of-Arc Programming

1. 圆心编程

 **General Syntax**

G2 Xn Zn In Kn Fn

or

G3 Xn Zn In Kn Fn

G2 Specifies the circular interpolation in the clockwise direction. See Figure 2-27.

G3 Specifies the circular interpolation in the counterclockwise direction. See Figure 2-28.

Xn Zn Specifies the X and Z absolute coordinates of the arc end point.

In Kn Specifies the incremental X and Z distances with + or - direction from the arc start point to the arc center.

Fn Specifies the feed rate of the tool. If not programmed, the system will use the last programmed feed rate.

 常用格式

G2 Xn Zn In Kn Fn

或

G3 Xn Zn In Kn Fn

G2 指顺时针方向的圆弧插补, 如图 2-27 所示。

G3 指逆时针方向的圆弧插补, 如图 2-28 所示。

Xn Zn 指定圆弧终点的 X 和 Z 绝对坐标。

In Kn 指从圆弧起点到圆弧中心的 X 向和 Z 向的矢量值。

Fn 指刀具的进给量。如果未编程,系统将使用程序中前面最近一次指定的数值。

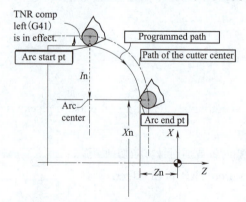

Figure 2-27　The Clockwise Direction

图 2-27　顺时针方向的圆弧插补

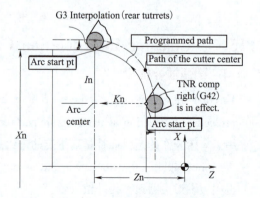

Figure 2-28　The Counterclockwise Direction

图 2-28　逆时针方向的圆弧插补

2. Radius-of-Arc Programming（for arcs less than or equal to 180°）
2. 圆弧半径编程（适用于圆心角小于或等于180°的弧）

G2 Xn Zn Rn Fn

or

G3 Xn Zn Rn Fn

G2 Specifies the circular interpolation in the clockwise direction. See Figure 2-29.

G3 specifies the circular interpolation in the counterclockwise direction. See Figure 2-30.

Xn Zn Specifies the X and Z absolute coordinates of the arc end point.

Rn Specifies the radius of the arc to be cut.

Fn Specifies the feed rate of the tool. If not programmed, the system will use the last programmed feed rate.

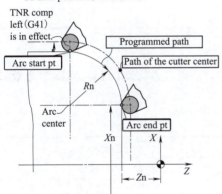

Figure 2-29 The Clockwise Direction

图 2-29 顺时针方向的圆弧插补

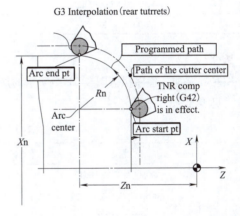

Figure 2-30 The Counterclockwise Direction

图 2-30 逆时针方向的圆弧插补

G2 Xn Zn Rn Fn

或

G3 Xn Zn Rn Fn

G2 指顺时针方向的圆弧插补，如图 2-29 所示。

G3 指逆时针方向的圆弧插补，如图 2-30 所示。

Xn Zn 指圆弧终点 X 和 Z 绝对坐标。

Rn 指切削的圆弧的半径。

Fn 指刀具的进给量。如果未编程，系统将使用程序中前面最近一次指定的数值。

▌EXAMPLE 2-12

Write blocks to execute the machining operations shown in Figure 2-31. Assume TNR compensation right（G42）is in effect. Word address Commands and the meanings See Table 2-9.

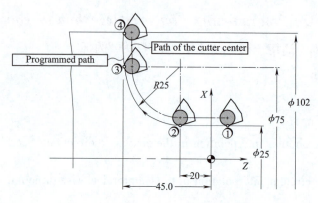

Figure 2-31　Case 2-12

图 2-31　例 2-12

Table 2-9　Word Address Commands and the Meanings

Word address command	Meaning
N0050 G1 Z-20 F0.12	Cut to ②
N0060 G2 X75 Z-45 R25	Cut R25 arc to ③
N0070 G1 X102	Cut to ④

■ 例 2-12

编写数控加工程序执行图 2-31 中所示的加工操作。假设刀尖圆弧半径补偿（G42）有效指令及其含义见表 2-9。

表 2-9　指令及其含义

指令	含义
N0050 G1 Z-20 F0.12	切到 ②
N0060 G2 X75 Z-45R25	将 R25 弧切到 ③
N0070 G1 X102	切到 ④

■ EXAMPLE 2-13

Code blocks to machine the profile shown in Figure 2-32. Assume TNR compensation right (G42) is in effect. Word address Commands and the meanings. See in Table 2-10.

■ 例 2-13

编写加工图 2-32 所示轮廓的程序段。假设刀尖圆弧半径补偿（G42）有效。指令及其含义见表 2-10。

Table 2-10　Word Address Commands and the Meanings

Word address command	Meaning
N0050 G1 Z-20 F0.12	Cut to ②
N0060 X25	Cut to ③
N0060 G3 X50 Z-33R13	Cut R25 arc to ④
N0070 G1 Z-38	Cut to ⑤

表 2-10 指令及其含义

指令	含义
N0050 G1 Z-20 F0.12	切到②
N0060 X25	切到③
N0060 G3 X50 Z-33R13	将 R25 弧切成④
N0070 G1 Z-38	切到⑤

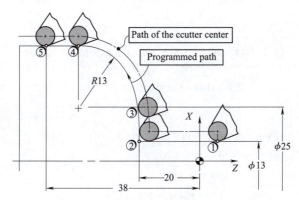

Figure 2-32　Case 2-13

图 2-32　例 2-13

■ **EXAMPLE 2-14**

The part shown in Figure 2-33 has been rough turned. Use TNR compensation to execute the facing and turning finish pass. Assume the following information has been entered into control's offset file during setup.

(1) Tool offset values for tool edge programming.

(2) Tool nose radius value.

(3) Tool nose vector 3.

Word address Commands and meanings. See Table 2-11.

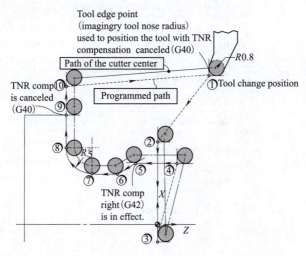

Figure 2-33　Case 2-14

图 2-33　例 2-14

例 2-14

图 2-33 中所示的零件已经完成粗加工。使用刀尖圆弧半径补偿切削端面和精车。假设在安装期间已将以下信息输入到机床的补偿文件中。

（1）刀尖编程的刀具补偿值。

（2）刀尖圆弧半径值。

（3）刀尖矢量 3。

指令及其含义见表 2-11。

Table 2-11　Word Address Commands and Meanings

Programming pattern	Word address command	Meaning
Job setup data listing	O1910	Main program number
	(X0 IS ON THE SPINDLE CENTERLINE)	Optional setup data
	(Z0 IS ON THE BLANK FACE)	
	(TOOLING LIST)	
	(TOOL 1:.8 TNR FINISHING TOOL)	
Machine start-up sequence Change to tool 1	N0010 G21	Metric mode
	N0020 G50 X203 Z101	Set machining origin at ①
	N0030 T0101	Change to tool 1
	N0040 G0 M42	Shift to intermediate gear range. Rapid on
	N0050 G96 S500 M3	Spindle on (CW) at 500 m/min, constant surface speed
Operation1：finish turn the end face	N0060 X76 Z0 T0101 M08	Rapid move to ②. Use tool edge control point when tool moves to ②. Coolant on
	N0070 G1 X-3 F0.15	Cut to ③ at feed rate
Operation2：finish turn the contour	N0080 G0 X68 Z2	Rapid move to ④. Tool edge programming is still in effect
	N0090 G1 G42 Z-20 F0.25	G42 (comp on) right. Move to ⑤ at feed rate. Tool will be offset to the right of its movement to ⑤. Use value in offset file 1
	N0100 X63 Z-2.5	Cut to ⑥
	N0110 Z-40	Cut to ⑦
	N0120 G2 X88 Z-53 R12	Cut 5R arc to ⑧
	N0130 G1 G40 X127	G40 (comp off) right. Move to ⑨ at feed rate. Use tool edge as control point when tool moves to ⑨
Machine stop. Program end sequence	N0140 G0 X132	Rapid to ⑩
	N0150 X203 Z101 T0100 M9	Rapid to tool change position ⑪. Cancel value in offset file 1. Coolant off
	N0160 M5	Spindle off
	N0170 M30	Program end. Memory reset

表 2-11 指令及其含义

编程内容	指令	含义
加工设置数据表	O1910	主程序号
	（X0 在主轴中心线上）	可选的设置数据
	（Z0 在端面面上）	
	（刀具清单）	
	（刀具 1：0.8TNR 精加工刀具）	
机床启动顺序 换 1 号刀	N0010 G21	米制模式
	N0020 G50 X203Z101	在①处设置机床原点
	N0030 T0101	更换为 1 号刀
	N0040 G0 M42	将主轴转到中速挡
	N0050 G96 S500 M3	以 500 m/min 主轴正转
操作 1： 完成端面加工	N0060 X76 Z0 T0101 M08	快速转移到②，当移动到②时使用刀具刀尖编程。打开冷却液
	N0070 G1 X-3F0.15	切到③
操作 2： 完成轮廓加工	N0080 G0 X68Z2.5	快速移动到④，刀尖编程仍然有效
	N0090 G1 G42 Z-20F0.25	刀尖圆弧半径右补偿，按进给量刀具移动⑤，刀具将向右移动到⑤，使用补偿文件 1 中的值
	N0100 X63Z-2.5	切到⑥
	N0110 Z-40	切到⑦
	N0120 G2 X88Z-53R12	将 R2.5 弧切到⑧
	N0130 G1 G40 X127	取消刀尖圆弧半径补偿，按进给量刀具移动转到⑨，在⑨点时，使用刀尖作为控制点
机床停止。 程序结束。	N0140 G0 X132	快速移动到⑩
	N0150 X203 Z101T0100 M9	快速移动到位置①，取消补偿文件 1 中数值，关闭冷却液
	N0160 M5	主轴关闭
	N0170 M30	程序结束，内存重置

■ **EXAMPLE 2-15**

The part shown (Figure 2-34) below has been previously roughed. Write a CNC program segment using TNR to finish bore the profile. The following information was entered into the control's offset file during setup.

(1) Tool offset values for tool edge programming.

(2) Tool nose radius value.

(3) Tool nose vector 2.

Word address Commands and meanings See Table 2-12.

■ 例 2-15

图 2-34 中所示的零件已完成粗加工。用刀尖圆弧半径补偿编写数控程序段完成精镗。下列信息被输入到补偿文件中。

（1）刀尖编程的刀具补偿值。

(2)刀尖圆弧半径值。

(3)刀尖矢量2。

指令及其含义见表2-12。

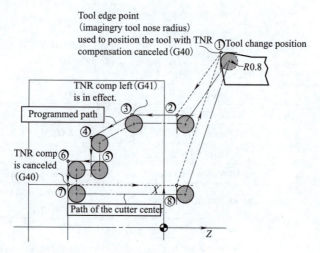

Figure 2-34　Case 2-15

图 2-34　例 2-15

Table 2-12　Word Address Commands and Meanings

Programming pattern	Word address command	Meaning
Job setup data listing	O1911	Main program number
	(X0 IS ON THE SPINDLE CENTERLINE)	Optional setup data
	(Z0 IS ON THE BLANK FACE)	
	(TOOLING LIST)	
	(TOOL 1:.8 TNR FINISHING TOOL)	
Machine start-up sequence Change to tool 1	N0010 G21	Metric mode
	N0020 G50 X304 Z127	Set machining origin at ①
	N0030 T0101	Change to tool 1
	N0040 G0 M42	Shift to intermediate gear range. Rapid on
	N0050 G96 S500 M3	Spindle on (CW) at 500 m/min, constant surface speed
	N0060 X66 Z2.5T0101 M08	Rapid move to ②. Use tool edge control point when tool moves to ②. Coolant on
Operation 1: finish turn the contour	N0070 G1 G41 Z-22F0.3	G41 (comp on) left. Move to ③ at feed rate. Tool will be offset to the left of its movement to ③. Use value in offset file 1
	N0080 G0 X55 Z-45	Cut to ④
	N0090 X45	Cut to ⑤
	N0100 Z-60	Cut to ⑥
	N0110 G40 X23	G40 (comp off). Move to ⑦ at feed rate. Use tool edge as control point when tool moves to ⑦

续表

Programming pattern	Word address command	Meaning
Machine stop. Program end sequence	N0120 G0 Z2.5	Rapid to ⑧
	N0130 X304Z127 T0100 M9	Rapid to tool change position ①. Cancel value in offset file 1. Coolant off
	N0140 M5	Spindle off
	N0150 M30	Program end. Memory reset

表 2-12 指令及其含义

编程内容	指令	含义
加工设置数据表	O1911	主程序号
	（X0 在主轴中心线上）	可选的设置数据
	（Z0 在零件端面上）	
	（刀具清单）	
	（刀具 1:0.8 TNR 精加工刀具）	
加工启动顺序 换 1 号刀	N0010 G21	米制模式
	N0020 G50 X304 Z127	在①处设置机床原点
	N0030 T0101	换 1 号刀
	N0040 G0 M42	将主轴转到中速挡
	N0050 G96 S500 M3	以 500 m/min 主轴正转
	N0060 X66Z2.5 T0101 M08	快速转移到②，当移动到②时使用刀具刀尖编程。打开冷却液
操作 1： 完成轮廓加工	N0070 G1 G41 Z-22F0.30	创建刀尖圆弧半径左补偿，按进给量移动到③，刀具将向左移动到③时。使用补偿文件 1 中的数值
	N0080 G0 X55 Z-45	切到④
	N0090 X45	切到⑤
	N0100 Z-60	切到⑥
	N0110 G40 X23	取消刀尖圆弧半径补偿，按进给量移动到⑦，当刀具移动到⑦时，使用刀尖作为控制点
机床停止。 程序结束。	N0120 G0 Z2.5	快速移动到⑧
	N0130 X304Z127 T0100 M9	快速移动至位置①，取消补偿文件 1 号位的值，关闭冷却液
	N0140 M5	主轴关闭
	N0150 M30	程序结束。内存重置

2.15.3 Grooving Commands

2.15.3 切槽指令

Grooving is executed by programming a linear cut with a specified dwell. The dwell is necessary to make the diameter of the groove uniform. The tool should be stopped at the bottom of the groove for at least one revolution of the spindle. See Figure 2-35.

通过使用指定的直线插补指令和特定的暂停来执行车槽。通过暂停可使槽的直径均匀，刀具停在槽底部的时间至少使主轴旋转一圈，如图 2-35 所示。

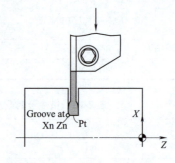

Figure 2-35　Grooving

图 2-35　切槽

General Syntax

G04 P*t*

G04 Specifies a programmed dwell.

P*t* The value of t sets the dwell time in milliseconds.

 常用格式

G04 P*t*

G04 指定暂停时间。

P*t* t 的值以毫秒为单位设置暂停时间。

 EXAMPLE 2-16

Code program blocks to machine the groove shown in Figure 2-36. Word address commands and the meanings. See Table 2-13.

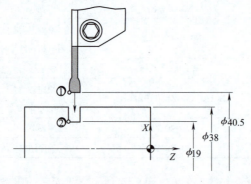

Figure 2-36　Case 9

图 2-36　例 9

Table 2-13　Word Address Commands and the Meanings

Word address command	Meaning
N0060 G0 X40	Rapid to①
N0070 G1 X19 F1.2	Cut to②at feed rate
N0080 G4 P500	Dwell for 500 millisecond
N0090 G0 X40	Rapid to①

 例 2-16

编写用于加工图 2-36 所示凹槽的程序段。指令及其含义见表 2-13。

表 2-13　指令及其含义

指令	含义
N0060 G0 X40	切到①
N0070 G1 X19 F1.2	按进给量切到②
N0080 G4 P500	暂停 500 ms
N0090 G0 X40	快速移动到①

2.16　Return to Reference Point Command

2.16　返回机床参考点指令

The CNC lathe can be programmed to automatically move the tool first to an intermediate point input and then to the reference point. These moves are made in rapid traverse.

可以对数控车床进行编程,使刀具首先自动快速移动到中间点,然后快速回到机床参考点。

General Syntax

G28 X*n* Z*n*

G28 Specifies a rapid move to the intermediate point and from there to the reference point.

X*n* Z*n* Specifies absolute coordinates of the intermediate point selected.

NOTES：

(1) The block G28 X0Z0 will cause a collision between the tool and workpiece and is never to be programmed.

(2) Because this command is used for automatic tool changing, cancel cutter diameter compensation and tool length compensation before coding a G28 block.

常用格式

G28 X*n* Z*n*

G28 指快速移动到中间点并回到参考点。

X*n* Z*n* 指中间点的绝对坐标。

注意事项：

（1）不要编写 G28 X0 Z0，否则将导致刀具与工件发生碰撞。

（2）由于此命令用于自动换刀，因此在编码 G28 程序段之前取消刀具直径补偿和刀具长度补偿值。

■ **EXAMPLE 2-17**

Code a block to automatically rapid the tool to the reference point by way of the intermediate point shown in Figure 2-37. Word address commands and the meanings See Table 2-14.

编辑一段程序，通过图 2-37 所示的中间点自动将刀具快速移动到参考点。

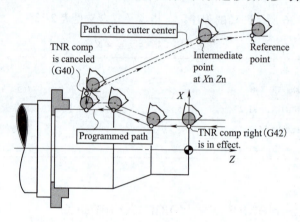

Figure 2-37　Case 2-17

图 2-37　例 2-17

Table 2-14　Word Address Commands and the Meanings

Word address command	Meaning
N0090 G28 X76 Z-12	Move the tool at rapid to the intermediate point at X76 Z-12 and from there to the reference point

■ **例 2-17**

编辑一段程序，通过图 2-37 所示的中间点并将刀具快速移动到参考点。指令及其含义见表 2-14。

表 2-14　指令及其含义

指令	含义
N0090 G28 X76 Z-12	将刀具快速移动到(76，-12)的中间点，然后回到参考点

Chapter Summary

本章总结

The following key concepts were discussed in this chapter.

（1）Lathe tools can be programmed to move in either incremental or absolute mode.

(2) With absolute diameter programming, the X position of the tool is specified in terms of diameter or twice the distance from the spindle centerline.

(3) The reference point (turret home position) is a point set once by the manufacturer.

(4) All tool movements are programmed relative to the part origin.

(5) The tool change point is located during setup and entered with a G50 block in the program.

(6) In small to intermediate shops the programmer prepares the setup sheet and the tool and operations sheet.

(7) Tool offsets for each tool are determined at setup and entered into the controller's memory.

(8) Lathe feed rates can be programmed in terms of G98 or G99.

(9) Automatic tool change is programmed as a Tab block, where a is two digit number of the turret station holding the tool, and b is a two digit number of the address in memory that contains the tool offset value.

(10) Tool nose radius (TNR) compensation programming involves programming the part geometry directly instead of the center of the tool nose.

(11) Since the controller automatically determines the location of the center of the tool nose at each point along a part boundary, mathematical computations are greatly simplified.

(12) A tool nose radius (TNR) compensation program can be written for a variety of tool nose radius sizes and to accommodate for tool wear.

(13) The word address command for tool nose radius compensation left of upward tool motion is G41. Compensation right of upward tool motion is G42 and compensation cancel is G40.

(14) Compensation is always applied by the controller in a direction perpendicular to the next in-plane axis move.

(15) For lathes, linear interpolation at feed rate is programmed with a G1 code.

(16) For lathes, Circular interpolation at federate is programmed with a G2 (CW) or G3 (CCW) word. If I and K are used, the arc center is specified. If R is used, the arc radius is indicated.

(17) Grooving involves programming a dwell (G4) after a linear interpolation (G1) block move.

(18) Machining at constant surface speed is important to maintaining optimum tool performance. A G96 Sn block signals the controller to continuously adjust the spindle RPM to achieve constant surface speed (Sn).

本章讨论了以下重要概念。

(1) 车床刀具可用增量编程或绝对程序模块。

(2) 使用绝对直径编程时,刀具的X轴位置以直径或距主轴中心线的距离的两倍来规定。

(3) 机床参考点是制造商设置的一个点。

(4) 所有刀具运动都相对于编程原点编程。

(5) 换刀点在机床操作过程中确定,并在程序中使用G50程序段输入。

(6)在中小型工厂中,程序员准备零件夹具图、刀具表和加工顺序表。

(7)机床操作阶段确定每个刀具的刀具补偿值,并将其输入控制器。

(8)车床进给量可以用 G98 或 G99 来编程。

(9)使用 Tab 指令进行自动换刀,其中 a 是刀架中夹持刀具的位置号,b 是包含刀具补偿值的两位数字的内存地址。

(10)刀尖圆弧半径补偿编程是指直接编程零件几何形状而不是刀尖圆角圆心位置。

(11)由于控制器自动确定沿零件轮廓的每个点处刀尖圆角中心的位置,因此大大简化了数学计算。

(12)刀尖圆弧半径补偿程序可针对各种刀尖半径尺寸编写,以适应刀具磨损。

(13)刀尖圆弧半径左补偿的指令为 G41。刀尖圆弧半径右补偿的指令为 G42,补偿取消指令为 G40。

(14)控制器总是在垂直于下一个平面轴向上施加补偿。

(15)对于车床,用 G1 代码编程进行直线插补编程。

(16)对于车床,圆弧插补用 G2(CW) 或 G3(CCW) 编程。如果使用 I 和 K,则指定圆弧中心。如果使用 R,则指定圆弧半径。

(17)切槽时,使用 G1 直线插补指令后暂停一段时间。

(18)以恒线速度加工对于保持最佳刀具性能非常重要。G96 Sn 指令控制控制器连续调节主轴转速以获得恒定的表面速度(Sn)。

Review Exercises

回顾练习

2.1 Write absolute coordinates for the points shown in Figure 2-38 in Table 2-15 (*X is expressed in terms of the diameter*).

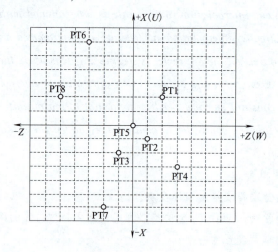

Figure 2-38　Case 11

图 2-38　例 11

Table 2-15   Absolute X and Z coordinates

POINT	X	Z
1		
2		
3		
4		
5		
6		
7		
8		

2.1 在表 2-15 中写出图 2-38 中点的绝对坐标(X 用直径表示)。

表 2-15 X 和 Z 绝对坐标

点	X	Z
1		
2		
3		
4		
5		
6		
7		
8		

2.2 Write incremental X and Z coordinates of the points in Table 2-16 shown in Figure 2-38. Use the following order: origin to PT1, from PT1 to PT2 from PT2 to PT3… Finish with PT8 (U is expressed in terms of the incremental diameter).

Table 2-16   Incremental U and W coordinates

Point	X	Z
1		
2		
3		
4		
5		
6		
7		
8		

2.2 在表 2-16 中写出图 2-38 所示点的 X 轴和 Z 轴增量坐标。按以下顺序：原点到 PT1，从 PT1 到 PT2 从 PT2 到 PT3 ……到 PT8 结束(U 用增量直径表示)。

表 2-16 增量 U 和 W 坐标

点	X	Z
1		
2		
3		
4		
5		
6		
7		
8		

2.3 For the profile (Figure 2-39) shown below, write the blocks for executing the tool movements at feed rate from points ③ to ④ to ⑤ in absolute (X, Y) and incremental (U, W) coordinates in Table 2-17. TNR comp right is in effect at these points and the feed rate is 0.3 mm/r.

Program number: O1914.

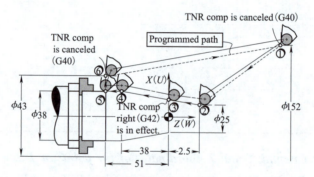

Figure 2-39 Case 2.3

图 2-39 例 2.3

Notes: Use 50 dia bar stock * 60 long and faced.

Table 2-17 Word Address Commands

Absolute coordinates	Incremental coordinates	Position
G1 G42 X Z F	G1 G42 U W F	③
X Z	U W	④
X Z	U W	⑤

2.3 对于图 2-39 显示的曲线,以绝对 (X,Y) 和增量 (U,W) 坐标从点③到④到⑤编写刀具运动的程序段并写入表 2-17 中。可以使用刀尖圆弧半径右补偿,进给量为 0.3 mm/r。

程序号:O1914。

注意事项:使用直径 50 mm,长 60 mm 的棒料,并且端面已经加工。

表2-17 程序

绝对坐标	增量坐标	位置
G1 G42 X Z F	G1 G42 U W F	③
X Z	U W	④
X Z	U W	⑤

2.4 The profile shown in Figure 2-40 is to be turned. Write blocks for executing the tool movements listed in both absolute and incremental modes and write in Table 2-18. Use a feed rate of 0.25 mm/r and a spindle speed of 500 r/min. Assume TNR right is in effect and the TNR is 0.3mm.

Program number: O1915.

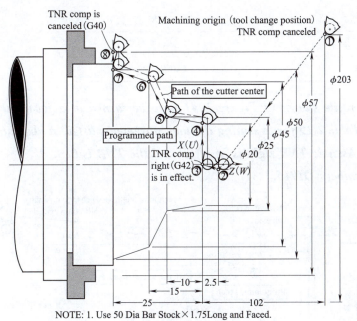

NOTE: 1. Use 50 Dia Bar Stock×1.75Long and Faced.

Figure 2-40 Case 13

图2-40 例13

Notes: Use 210 dia bar stock * 150 long and faced.

Table 2-18 Word Address Commands

Absolute coordinates	Incremental coordinates	Position
G1 G42 X Z F	G1 G42 U W F	③
X Z	U W	④
X Z	U W	⑤
X Z	U W	⑥
X Z	U W	⑦
G40 X Z	G40 U W	⑧

2.4 车削图2-40中所示的轮廓。以绝对和增量坐标编写数控加工程序并写在表2-18中。使用0.25 mm/r的进给量和500 r/min的主轴转速。假设刀尖圆弧半径右补偿有效,刀尖圆弧半径为0.3 mm。

程序号：O1915。

注意事项：使用直径210 mm,长150 mm的棒料,并且端面已经加工。

表 2-18 程序

绝对坐标	增量坐标	位置
G1 G42 X Z F	G1 G42 U W F	③
X Z	U W	④
X Z	U W	⑤
X Z	U W	⑥
X Z	U W	⑦
G40 X Z	G40 U W	⑧

2.5 The profile shown in Figure 2-41 is to be turned at a feed rate of 0.1 mm/r. Write blocks in Table 2-19 for executing the tool movements listed in absolute and incremental coordinates. Assume TNR right is in effect and the TNR is 0.3.

Program number：O1916.

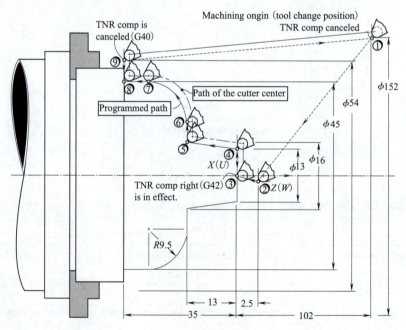

Figure 2-41　Case 2.5
图 2-41　例 2.5

Notes：Use 160 dia bar stock ＊ 150 long and faced.

Table 2-19　Word Address Commands

Absolute coordinates	Incremental coordinates	Position
G1 G42 X Z F	G1 G42 U W F	③
X Z	U W	④
X Z	U W	⑤
X Z	U W	⑥
G3 X Z I K	G3 U W I K	⑦
G1 X Z	G1 U W	⑧
G40 X Z	G40 U W	⑨

2.5　刀具以 0.1 mm/r 的进给量切削图 2-41 所示的轮廓。在表 2-19 中写入程序段，按绝对坐标和增量坐标列出的刀具运动。假设刀尖圆弧半径右补偿有效，刀尖圆弧半径为 0.3 mm。

程序号：O1916。

注意事项：使用直径 160 mm，长 150 mm 的棒料，并且端面已经加工。

表 2-19　程序

绝对坐标	增量坐标	位置
G1 G42 X Z F	G1 G42 U W F	③
X Z	U W	④
X Z	U W	⑤
X Z	U W	⑥
G3 X Z I K	G3 U W I K	⑦
G1 X Z	G1 U W	⑧
G40 X Z	G40 U W	⑨

2.6　Explain the significance of the following points.

a. Reference point　　b. Machining origin　　c. The program origin

d. Tool change position

2.6　解释以下几点的重要性。

a. 机床参考点　　b. 机床原点　　c. 编程原点　　d. 换刀点

2.7　a. What are tool offsets with regard to CNC lathes?

b. How are tool offsets used to execute a programmed tool move?

2.7　a. 什么是数控车床的刀具补偿？

b. 如何使用刀具补偿编写刀具加工程序？

Chapter 3　Techniques and Fixed Cycles for CNC Lathe Programming

第 3 章　车削固定循环指令

Learning Objectives

At the conclusion of this chapter you will be able to:

(1) Explain and program the rough boring cycle G90 and rough facing cycle G94.

(2) Understand and program the stock removal in turning and boring cycle G71.

(3) Know and program the finish turning and boring cycle G70.

(4) Describe and program the peck drilling and face grooving cycle G74.

(5) Understand and program the peck cutoff and grooving cycle G75.

(6) Describe and program the following threading cycles: single pass G32, multiple pass G92, and multiple repetitive G76.

学习目标

在本章结束时,你将能够:

(1)解释并使用粗加工轮廓循环指令 G90 和端面粗加工循环指令 G94。

(2)了解和使用车削和镗孔循环指令 G71 中的切削量。

(3)了解并使用精加工车削和镗孔循环指令 G70。

(4)描述和使用钻孔和端面切槽循环指令 G74。

(5)了解并使用切断和切槽循环指令 G75。

(6)描述并使用以下螺纹循环指令:单线循环指令 G32、多线循环指令 G92 和复合循环指令 G76。

CNC lathe machining cycles make programming operations easier and more efficient. Cycles for rough turning, boring, and facing straight or tapered profiles are presented first, followed by a discussion of cycles for stock removal in turning and boring as well as finishing. Grooving is an important lathe operation executed at the end of a thread to ensure full engagement of the nut up to the shoulder. It is also performed at the edge of a shoulder to permit the proper fit of mating parts. Cycles for peck drilling and grooving are considered in detail. The chapter ends with complete descriptions of the most important CNC lathe threading cycles.

数控车床加工循环指令使编程操作更简便、有效。首先介绍了粗加工、镗孔、车削平端面或斜端面循环指令,然后讨论了车削和镗削的精加工中切削循环指令。车槽是在螺纹末端的重要车床操作，以确保螺纹完全啮合到底。它也在顶端的边缘进行，使部件更好地进行配合。其次，详细介绍了钻孔和切槽的循环指令。本章最后还介绍了最重要的数控车床螺纹加工循环指令。

There are some safety rules for programming CNC lathes, they are：

(1) Always check for any interference between the work piece, the tooling setups, and the indexing of each turret face.

(2) Always select cutters, holders, and turret accessories that provide maximum rigidity for holding during machining.

(3) Select speeds and feeds that permit machining at optimum efficiency and safety.

(4) Reduce the speed and feed：

- When drilling large-diameter holes to a depth greater than twice the drill size.
- When cutting thread.

(5) When finished cutting an internal tape, always feed the tool in the direction of the large diameter.

(6) During rough turning, pay attention to the following：

- The effect of longitudinal, tangential, and radial forces.
- The heat generated between the work piece and tool.

数控车床编程时，需要遵循一些原则：

(1) 始终检查工件、工装面和刀具换刀面之间是否有任何干扰。

(2) 始终确保刀具、刀柄和刀架附件在加工过程中提供最大的刚性。

(3) 选择最高效和最安全的切削速度和进给量进行加工。

(4) 降低切削速度和进给量：

- 当钻孔深度大于钻头直径时。
- 切削螺纹时。

(5) 完成工件内部切削后，始终按大直径方向进给刀具。

(6) 在粗加工期间，请注意以下事项：

- 纵向、切向和径向力的影响。
- 工件和刀具之间产生的热量。

3.1　Turning and Boring Cycle：G90

3.1　外圆车削和镗孔循环指令：G90

A single block containing a G90 word executes repetitive straight-cut machining passes required to turn or bore a part from stock.

G90 指令可以完成车削外圆或内孔的一系列直线运动。

G90轴向单一循环指令（圆柱切削）2D轨迹

📝 General Syntax

G90 Xa Za Fn

Xb Za

⋮

Xn za

G90 signals the controller to begin the straight-cut turning or boring cycle. See Figure 3-1.

Xa Za，Xb Za，⋯Xn Za are the absolute coordinates of the tool at the end of each machining pass.

Fn is the value of n specifies the feed rate.

G90矩形车削循环指令3D视频

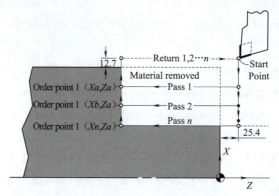

Figure 3-1　The Cutting Path of Gpo

图 3-1　指令的加工路径

📝 常用格式

G90 Xa Za Fn

Xb Za

⋮

Xn Za

G90 向控制器发出信号以直线切削完成车削或镗孔循环，如图 3-1 所示。

Xa Za，Xb Za,⋯,Xn Za 刀具在切削终点的绝对坐标。

G90轴向单一循环指令（圆锥切削）2D轨迹

Fn n 指进给量的值。

NOTES：

（1）Position the tool at the start before entering the G90 command.

（2）For incremental coordinated replace X with U, and Z with W. U and W indicate the direction and distance from the start point to the order point along the X and Z axes, respectively.

（3）The blocks G90 Xa Za R Fn

Xb Za

⋮

Xn Za

can be used to execute taper cutting in the direction along the spindle axis. The R address replaces the older I address. R or I specifies the height of the taper and the direction (+ or −) the tool moves in making the first cut for the taper. See Figure 3-2.

$$R \text{ or } L = \frac{D_{start} - D_{end}}{2}$$

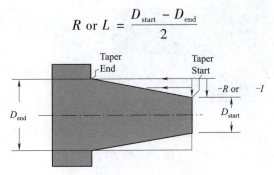

Figure 3-2　Processing Example

图3-2　加工实例

注意事项：

（1）在输入 G90 命令之前，将刀具运动到循环起始点。

（2）对于增量编程，用 U 代替 X，用 W 代替 Z。U 和 W 分别表示沿 X 轴和 Z 轴从起点到切削终点的方向和距离。

（3）指令段 G90 Xa Za R Fn

Xb Za

⋮

Xn Za

可用于沿主轴轴线方向锥形切削。R 地址替换旧的 I 地址。R 或 I 指定锥度的高度以及刀具在进行锥度的第一次切削时的移动方向（＋或−），如图3-2所示。

$$锥度的高度 = \frac{切削起点直径 - 切削终点直径}{2}$$

■ **EXAMPLE 3-1**

Write a program segment for rough turning the part, as illustrated in Figure 3-3. Use the G90 cycle and assume all information needed for programming the tool edge has been entered into the controller's offset file.

Because the profile consists only of horizontal and vertical lines, tool edge programming may be directly applied. Word address command and meaning see Table 3-1.

■ **例3-1**

编写程序段进行粗加工，如图3-3所示。使用 G90 循环并假设编程刀具刀尖所需的所有信息都已输入到控制器的补偿文件中。

由于工件轮廓仅由水平和垂直线组成,因此可以直接应用刀具刀尖编程。指令及其含义见表 3-1。

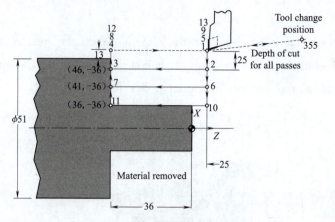

Figure 3-3　Case 3-1

图 3-3　例 3-1

Table 3-1　Word Address Command and Meanings

Programming pattern	Word address command	Meaning
—	O2001	Program number
	(TOOL 1:.8 TNR FINISHING TOOL)	Optional setup data
	(ROUGH BORE THE PART)	
Machine start-up sequence. Change to tool 1	N0010 G21	Metric mode
	N0020 G50 X203 Z101	Set machining origin at 14
	N0030 G50 S3000	Set spindle RPM limit to 3000 r/min
	N0040 T0100	Return to tool change position 14. Index to tool 1 Cancel values in offset file 1
	N0050 G0 M42	Shift to intermediate gear range. Rapid on
	N0060 G96 S500M3	Spindle on (CW) at 500SMM, constant surface speed
	N0070 X53 Z2.5 T0101 M8	Rapid move to ①。Use value in offset file 1. Coolant on
Operation 1:finish turn the contour	N0080 G90 X45 Z-36 F0.25	Excute multiple-pass straight-cutting cycle: Rapid to ②, cut to ③ and ④ at feed rate, rapid to ⑤
	N0090 X40	Rapid move to ⑥. Cut to ⑦ and ⑧ at feed rate, rapid to ⑨
	N0100 X35	Rapid move to ⑩. Cut to 11 and 12 at feed rate, rapid to 13
Machine stop. Program end sequence	N0110 G0 X203 Z101 T0100 M9	Rapid to tool change position14. Cancel value in offset file 1. Coolant off
	N0120 M5	Spindle off
	N0130 M30	Program end. Memory reset

表 3-1 指令及其含义

程序模块	指令	含义
—	O2001	程序编号
	（刀具 1：0.8 TNR 精加工刀具）	可选的设置数据
	（粗加工部分）	
加工启动顺序 换 1 号刀	N0010 G21	米制模式
	N0020 G50 X203 Z101	将机床原点设置为 14
	N0030 G50 S3000	将主轴最大转速设置为 3 000 r/min
	N0040 T0100	返回换刀位置 14，换 1 号刀 1，取消补偿文件 1 中的值
	N0050 G0 M42	将主轴转到中速挡
	N0060 G96 S500M3	主轴 500 m/min 正转
	N0070 X53 Z2.5 T0101 M8	快速移动到①，使用补偿文件 1 中的值，打开冷却液
操作 1： 完成轮廓加工	N0080 G90 X45 Z-35 F0.25	执行 G90 循环：快速到②，以进给量切削到③和④，快速移动到⑤
	N0090 X40	快速移动到⑥，以进给量切削到⑦和⑧，快速移动到⑨
	N0100 X35	快速移动到⑩，以进给量切削到 11 和 12，快速移动到 13
加工停止 程序结束	N0110 G0 X203 Z101 T0100 M9	快速移动到位置 14，取消补偿文件 1 中的值。关闭冷却液
	N0120 M5	主轴关闭
	N0130 M30	程序结束。内存重置

■ **EXAMPLE 3-2**

Use the G90 cycle to write a program segment for rough boring the part shown in Figure 3-4. Assume all the information for tool edge programming has been entered into the controller's offset file.

Tool edge programming may be directly applied for executing the horizontal and vertical line cuts. Word address command and meaning see Table 3-2.

■ **例 3-2**

使用 G90 循环编写程序段，用于粗加工图 3-4 中所示的零件。假设刀具刀尖编程的所有信息都已输入到控制器的补偿文件中。

可以直接应用刀具刀尖编程来进行水平和直线切削。指令及其含义见表 3-2。

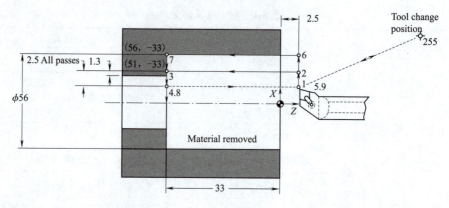

Figure 3-4 Case 3-2

图 3-4 例 3-2

Table 3-2 Word Address Command and Meanings

Programming pattern	Word address command	Meaning
—	O2002	Program number
	(TOOL 1:.8 TNR FINISHING TOOL)	Optional setup data
	(ROUGH BORE THE PART)	
Machine start-up sequence. Change to tool 1	N0010 G21	Metric mode
	N0020 G50 X203 Z101	Set machining origin at ⑩
	N0030 G50 S3000	Set spindle rpm limit to 3 000 r/min.
	N0040 T0100	Return to tool change position ⑩. Index to tool 1. Cancel values in offset file 1
	N0050 G0 M42	Shift to intermediate gear range. Rapid on
	N0060 G96 S600 M3	Spindle on (CW) at 600 m/min, constant surface speed
	N0070 X45 Z2.5 T0101 M8	Rapid move to ①. Use values in offset file 1. Coolant on
Operation 1:finish turn the contour	N0080 G90 X50 Z-33 F0.3	Excute multiple-pass straight-cutting cycle: Rapid to ②, cut to ③ and ④ at feed rate, rapid to ⑤
	N0090 X55	Rapid move to ⑥. Cut to ⑦ and ⑧ at feed rate, rapid to ⑨
Machine stop. Program end sequence	N0100 G0 X203 Z201 T0100 M9	Rapid to tool change position ⑩. Cancel value in offset file 1. Coolant off
	N0110 M5	Spindle off
	N0120 M30	Program end. Memory reset

表3-2 指令及其含义

程序模块	指令	含义
—	02002	程序编号
	（刀具1:.8 TNR 精加工刀具）	可选的设置数据
	（粗加工部分）	
加工启动顺序 换1号刀	N0010 G21	米制模式
	N0020 G50 X203 Z101	将机床原点设置为⑩
	N0030 G50 S3000	将主轴最大转速设置为3 000 r/min
	N0040 T0100	返回换刀位置⑩，换1号刀，取消补偿文件1中的值
	N0050 G0 M42	将主轴转到中速挡
	N0060 G96 S600 M3	主轴600 m/min 速度正转
	N0070 X45 Z2.5 T0101 M8	快速移动到①，使用补偿文件中的值1，打开冷却液
操作1： 完成轮廓精加工	N0080 G90 X50 Z-33 F0.3	执行G90循环：快速到②，使用进给量切削到③和④，快速移动到⑤
	N0090 X55	快速移动到⑥，使用进给量切削到⑦和⑧，快速移动到⑨
加工停止。 程序结束	N0100 G0 X203 Z101 T0100 M9	快速至换刀位置⑩，取消补偿文件1中的值。关闭冷却液
	N0110 M5	主轴关闭
	N0120 M30	程序结束。内存重置

3.2 Facing Cycle G94

3.2 端面车削循环指令 G94

The G94 word enables the programmer to execute the multiple straight-cut machining passes needed to face a part from stock.

G94 使编程人员能够执行端面车削所需的多个直线切削加工过程。

General Syntax

G94 Xa Za Fn

Xa Zb

⋮

Xa Zn

G94 Directs the controller to begin the straight-cut facing cycle. See Figure 3-5.

Xa Za, Xa Zb,…,Xa Zn Are the absolute coordinates of the tool at the end of each straight-cut (order point).

Fn Value of n specifies the feed rate.

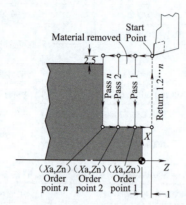

Figure 3-5　The Cutting Path of G94

图 3-5　G94 指令的加工路径

常用格式

G94 Xa Za Fn

Xb Zb

⋮

Xn Zn

G94 指示控制器开始端面车削循环,如图 3-5 所示。

Xa Za, Xa Zb,…,Xa Zn 刀具在切削终点的绝对坐标。

Fn n 指定进给量。

NOTES：

(1) Position the tool at the start before programming the G94 cycle.

(2) For incremental coordinates replace X with U, and Z with W. U and W indicate the direction and distance from the start point to the order point along the X and Z axes, respectively.

(3) The blocks G94 Xa Za R Fn

Xb Za

⋮

Xn Za

can be used to execute taper cutting in the direction perpendicular to tine-spindle axis. The R address replaces the older K address. R or K specifies the height of the taper along the spindle axis. See Figure 3-6.

注意事项：

(1)在编辑 G94 循环之前,先将刀具置于循环起始点。

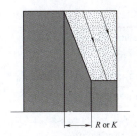

Figure 3-6　Processing Example

图 3-6　加工实例

(2) 对于增量坐标，将 X 替换为 U，将 Z 替换为 W，U 和 W 分别表示沿 X 轴和 Z 轴从循环起点到切削终点的方向和距离。

(3) 指令段 G94 Xa Za R Fn
　　　　　　Xb Za
　　　　　　⋮
　　　　　　Xn Za

可用于在垂直于主轴轴线的方向上的锥形切削。R 地址替换旧的 K 地址。R 或 K 指定沿主轴轴线的锥度高度，如图 3-6 所示。

■ **EXAMPLE 3-3**

Use the G94 cycle to write a program segment for rough facing the part shown in Figure 3-7. Assume all the information for tool edge programming has been entered into the controller's offset file.

Tool edge programming will, again, be used directly to execute the horizontal and vertical line cuts. Word address command and meanings see Table 3-3.

■ **例 3-3**

图 3-7 所示的工件端面需要进行粗加工。使用 G94 循环编写所需的程序段。假设刀具补偿文件包含刀具刃口编程的所有信息。

刀具刀尖编程将再次直接用于水平和垂直线切削。指令及其含义见表 3-3。

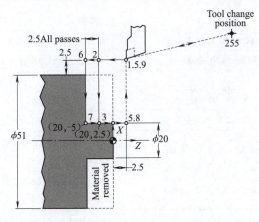

Figure 3-7　Case 3-3

图 3-7　例 3-3

视频

G94端面循环切削指令2D轨迹

Table 3-3　Word Address Command and Meanings

Programming pattern	Word address command	Meaning
—	02003	Program number
	(TOOL 1:0.8 TNR FINISHING TOOL)	Optional setup data
	(ROUGH BORE THE PART)	
Machine start-up sequence. Change to tool 1	N0010 G21	Metric mode
	N0020 G50 X203 Z101	Set machining origin at ⑩
	N0030 G50 S3000	Set spindle rpm limit to 3 000 r/min
	N0040 T0100	Return to tool change position ⑩. Index to tool 1. Cancel values in offset file 1
	N0050 G0 M42	Shift to intermediate gear range. Rapid on
	N0060 G96 S600 M3	Spindle on (CW) at 600 m/min, constant surface speed
	N0070 X56 Z2.5 T0101 M8	Rapid move to ①. Use values in offset file 1. Coolant on
Operation 1: finish turn the contour	N0080 G94 X20 Z-2.5 F0.30	Excute multiple-pass straight-cutting cycle: Rapid to ②, cut to ③ and ④ at feed rate, rapid to ⑤
	N0090 Z-5	Rapid move to ⑥. Cut to ⑦ and ⑧ at feed rate, rapid to ⑨
Machine stop. Program end sequence	N0100 G0 X203 Z101 T0100 M9	Rapid to tool change position ⑩. Cancel value in offset file 1. Coolant off
	N0110 M5	Spindle off
	N0120 M30	Program end. Memory reset

表 3-3　指令及其含义

程序模块	指令	含义
—	02003	程序编号
	（刀具 1:0.8 TNR 精加工刀具）	可选的设置数据
	（粗加工部分）	
加工启动顺序 换1号刀	N0010 G21	米制模式
	N0020 G50 X203 Z101	将机床原点设置为 ⑩
	N0030 G50 S3000	将主轴最大转速设置为 3000 r/min
	N0040 T0100	在位置 ⑩ 换 1 号刀，取消补偿文件 1 中的值
	N0050 G0 M42	将主轴转到中速挡
	N0060 G96 S600 M3	主轴以 600 m/min 速度正转
	N0070 X56 Z2.5 T0101 M8	快速移动到 ①，使用补偿文件 1 中的值，打开冷却液
操作 1： 完成轮廓加工	N0080 G94 X20 Z-2.5 F0.3	执行 G94 循环：快速到 ②，按进给量切削到 ③ 和 ④，快速移动到 ⑤
	N0090 Z-5	快速移动到 ⑥。按进给量切削到 ⑦ 和 ⑧，快速移动到 ⑨

续表

程序模块	指令	含义
加工停止，程序结束	N0100 G0 X203 Z101 T0100 M9	快速至位置⑩，取消补偿文件1中的值，关闭冷却液
	N0110 M5	主轴关闭
	N0120 M30	程序结束。内存重置

3.3　Multiple Repetitive Cycles：G70 TO G75

3.3　复合循环指令：G70～G75

The repetitive cycles (G70 to G75) reduce the effort involved in programming rough machining operations. These cycles simply require the programmer to specify the dimensions of the finished shape and cutting parameters. The controller responds by directing the tool to execute the necessary repetitive cutting. The tool will automatically remove material from stock so that the finish shape is achieved.

重复循环（G70～G75）减少了编程粗加工操作所需的工作量。这些循环只需要程序员指定零件最终形状和切削参数的尺寸。控制器通过控制刀具执行必要的重复切削来完成加工。该刀具将自动从毛坯上切除多余材料，从而加工成最终的形状。

3.3.1　Stock Removal in Turning and Boring Cycle：G71

3.3.1　粗车及粗镗循环指令：G71

The G71 cycle is used to rough cut a general contour from stock. The contour may consist of lines (horizontal, vertical, or tapered) and circular arcs. OD (turning) or ID (boring) operations may be programmed. The cutting path of G71 see Figure 3-8.

G71循环用于工件的轮廓粗加工。轮廓可以由直线（水平、垂直或锥形）和圆弧组成，可用于外圆或内孔加工的编程。G17指令的切削路径如图3-8所示。

> **General Syntax**

G71 Un Rn (BLOCK1)

G71 PNs QNf Un Wn Fn Sn (BLOCK 2)

NNs

⋮

NNf

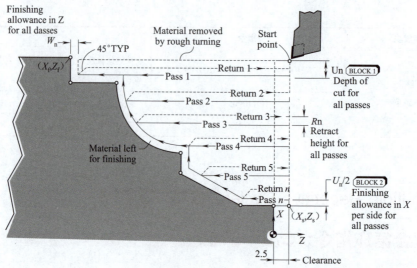

Figure 3-8 The Cutting Path of G71

图 3-8 G71 指令的切削路径

G71 Initiates rough cutting of the contour described in blocks Ns through Nf. See in Figure 3-8。

Un（BLOCK1）n, sets the depth of cut for each roughing pass. Decimal point not accepted.

Rn n sets the retract height for each roughing pass.

PNs Ns specifies the sequence number of the block which starts the contour description.

QNf Nf specifies the sequence number of the block which ends the contour description.

Un（BLOCK2）n Specifies direction the amount of finishing allowance along the X-axis（diameter values）.

Wn n specifies the direction amount of finishing allowance along the Z-axis

Fn n specifies the roughing feed rate.

Sn n specifies the roughing speed. Use G96 prior to G71 to set S to mpm（m/min）.

常用格式

G71 启动程序段粗加工从 Ns 到 Nf 的轮廓, 如图 3-8 所示。

Un（BLOCK1）n 设置每次粗加工的切削深度。

Rn n 为退刀量。

PNs Ns 指定轮廓加工开始的程序段的序列号。

QNf Nf 指定轮廓加工结束的程序段的序列号。

Un（BLOCK2）n 指定沿 X 轴的精加工余量（直径值）。

Wn n 指定沿 Z 轴的精加工余量。

Fn n 指定粗加工进给量。

Sn n 指定粗加工主轴转速, 在 G71 之前使用 G96 将 S 单位设置为 m/min。

NOTES:

（1）Linear interpolation（G1）, corner rounding, chamfering, and circular Interpolation（G2/G3）may be used to describe the contour in blocks P through Q.

(2) Any F and S functions contained in Blocks P through Q are ignored. The functions programmed just prior to or in the G71 block are effective throughout the cycle.

(3) For the finishing allowances specification Un (BLOCK2), the sign of n will vary as follows:
- n is + for OD mind.
- n is - for ID boing.

注意事项:

(1) 可以使用直线插补指令(G1)、倒圆角、倒直角和圆弧插补(G2/G3)来加工段 P 到 Q 中的轮廓。

(2) 忽略程序段 P 到 Q 中包含的任何 F 和 S 指令。在 G71 程序段之前或 G71 中编辑的任何 F 和 S 指令在整个循环中都有效。

(3) 对于精加工程序段 Un,将如下变化:
- 对于外圆切削,n 为正值。
- 对于内孔切削,n 为负值。

G70精车循环指令2D轨迹

G71轴向粗车复合循环指令2D轨迹

3.3.2 Finish Turning and Boring Cycle: G70
3.3.2 精车及精镗循环指令: G70

The G70 cycle can be used to execute finish passes over a contour that has been roughed by a G71 cycle.

G70 循环可用于在已经由 G71 循环粗加工的轮廓上进行精加工。

 General Syntax

G70 PsNQNf

G70 Initiates finish cutting the contour described in blocks Ns through Nf. See Figure 3-9.

G70精加工循环指令3D视频

 常用格式

G70 PsNQNf

G70 对程序段 Ns 到 Nf 的轮廓进行精加工,如图 3-9 所示。

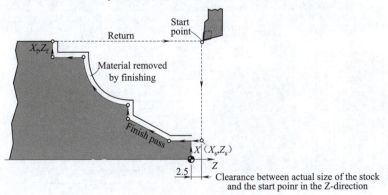

Figure 3-9 The Cutting Path of G70

图 3-9 G70 的加工路径

G71外径粗车循环指令3D视频

NOTES:

● 视频
G72端面
粗加工
循环指令
2D轨迹

(1) G70 can only be used after a G71 or G72 cycle within the same program.

(2) Any feed, F and speed, S address coded in blocks P through Q are automatically effective within the G70 cycle.

(3) Position the tool at the start point before issuing the G70 code.

注意事项：

(1) G70 只能在同一程序中的 G71 或 G72 循环后使用。

(2) 程序段 P 至 Q 定义的任何进给量 F、主轴转速 S 在 G70 循环内自动生效。

(3) 在启动 G70 代码之前将刀具定位在循环起始点。

● 视频
G72端面
车削固定
循环指令
3D视频

■ **EXAMPLE 3-4**

The part shown in Figure 3-10 is to be machined from stock. Write a program to execute the necessary roughing passes and the finish pass. The tool offsets for tool edge programming, tool nose radius, and tool nose vector 3 have been entered into the offset file.

Thus, the system will execute eight full-depth passes and one remaining stock pass.

$$\text{Number of rough passes} = \frac{\text{Stock diameter-Minor diameter of part} + \text{Material finish}}{\text{Depth of cut} * 2}$$

$$= \frac{114 - 45 + 0.6}{4 \times 2} = 9$$

■ **例 3-4**

加工如图 3-10 所示的零件。编写一段程序来执行粗加工和精加工。刀具刀尖编程补偿、刀尖圆弧半径和刀尖矢量 3 已输入到补偿文件中。

因此，机床将执行 8 次直线切削和 1 次轮廓切削。

$$\text{粗车循环次数} = \frac{\text{毛坯直径} - \text{零件上的最小直径} + \text{精车加工量}}{\text{切削深度} \times 2}$$

$$= \frac{114 - 45 + 0.6}{4 \times 2} = 9$$

● 视频
G73仿形
车削复合
循环指令
2D轨迹

G73固定
形状粗加工
循环指令
3D视频

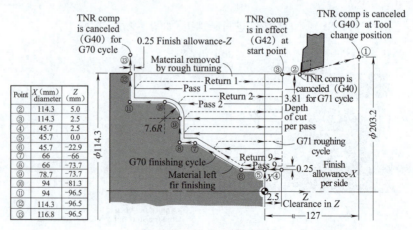

Figure 3-10　Case 3-4

图 3-10　例 3-4

CNC tool and cutting parameters used to rough turn OD and finish OD contour is shown in Tabel 3-4. And program to execute the necessary roughing passes and the finish pass is listed in Table 3-5 and Table 3-6. 粗车外圆和精车外圆的刀具和切削参数见表 3-4。加工程序在表 3-5 和表 3-6 中。

Table 3-4 CNC Tool and Operations Sheet

MATERIAL: AL ALLOY 7075 – T6

TOOL	OFF-SET	OPERATION	SPEED (m/min)	FEED (mm/r)	TOOL DESCRIPTION	INSERT	TNR
1	01	ROUGH TURN OD: ● USE A G71 ROUGHING CYCLE; ● REMOVE 3.81 PER SIDE FOR EACH PASS; ● LEAVE 0.25 FOR FINISHING	450	0.3	TURNING HOLDER MCLNR 12 – 4	CNMG – 432 – NR	0.8
2	02	FINISHING OD COUNTOUR: USE A G76 FINISHING CYCLE	800	0.13	TURNING HOLDER MCLNR 12 – 4	CNMG – 434 – NF	1.6

Table 3-5 Word Address Command and Meanings

Programming pattern	Word address command	Meaning
	O2004	
Job setup date listing	(..........) (X0 IS ON THE SPINDLE CENTERLINE) (Z0 IS ON THE BLANK FACE) (..........) (TOOLING LIST:) (TOOL 1:.8TNR ROUGHING TOOL) (TOOL 2:1.6TNR FINISHING TOOL) (..........)	Optional setup statements
Too l2 in turret Machine start-Up sequence	N0010 G21	Metric mode
	N0020 G50 X203 Z127	Set machining origin at ①
	N0030 G50 S3 000	Set spindle RPM limit to 3 000 r/min
	N0040 T0100	Return to tool change position①. Index to tool 1. Cancel values in offset file 1
	N0050 G0 M42	Shift to intermediate gear range
	N0060 G96 S450M3	Spindle on (CW) at 450 m/min, constant surface speed
	(TOOL 1:ROUGH TURN CONTOUR)	
	N0070 X114 Z5 T0101 M8	Rapid to②. Coolant on
TRN comp on(**G42**) prior to after of **G71** Cycle	N0080 G1 G42 Z2.5 F0.15	Ramp on to the right of upward tool motion on the next linear move to③ at feed 0.15

113

续表

Programming pattern	Word address command	Meaning
Operation 1: Rough turn contour with **G71** cycle	N0090 G71 U3.81	Start rough turning cycle. Cut to 3.81 mm Per side for each roughing pass
	N0100 G42 G71 P0100 Q0190 U0.5 W0 F0.3	Rough turn contour defined by blocks 0110 ~ 0190. Leave 0.25 mm Per side along the X axe. Machine at feed 0.3
	N0110 G0 X45.7	Rapid to ④ on contour
	N0120 G1 X45.7 Z0	Cut to ⑤ on contour
	N0130 Z−22.9	Cut to ⑥ on contour
	N0140 X66 Z−66	Cut to ⑦ on contour
	N0150 Z−73.7	Cut to ⑧ on contour
	N0160 X78.7	Cut to ⑨ on contour
	N0170 G3 X94 Z−81.3 R7.6	Cut R3 arc to ⑩ on contour
	N0180 G1 Z−96.5	Cut to11 on contour
	N0190 X114.3	Cut to12 on contour
	N0200 G0 Z2.5	Rapid to start point ③

Table 3-6 Word Address Command and Meanings

	Word address command	Meaning
—	（TOOL2：FINISH TURN CONTOUR）	
TNR comp off（**G40**）after **G71** cycle is completed	N0210 G1 G40 Z5	Ramp off on the next linear move to ② at feed rate
Machine stop sequence	N0220 G0 X203.2 Z127 T0100 M9	Rapid move to the tool change position ①. Cancel values in offset file 1. Coolant off
	N0230 M5	Spindle off
Change to tool2 Machine start-up sequence	N0240 T0200	Index to tool 2. Cancel values in offset register2
	N0250 G0 M43	Shift to high-gear range
	N0260 G96 S800 M3	Spindle on（CW）at 800 r/min, constant surface speed
	N0270 X114.3 Z5 T0202 M8	Rapid to ②. Coolant on
TNR comp on(**G42**) Prior to start of **G70** cycle	N0280 G1 G42 Z2.5 F0.1	Ramp on to the right of upward tool motion on the next linear move to ③ at feed 0.1
Operation 2: Finish turn contour with **G70** cycle	N0290 G42 G70 P0110 Q0190	Finish turn contour defined by blocks 0110 ~ 0190. Machine at feed 0.1 and speed 800 r/min
TNR comp off（**G40**）After **G70** cycle In completed	N0300 G1 G40 X116.8	Ramp off on the next linear move to12 at feed rate
Machine stop Program end sequence	N0310 G0 X203.2 Z127 T0200 M9	Rapid move to the tool change position ①. cancel values in offset file 2. coolant off
	N0320 M5	Spindle off
	N0330 M3	Program end. Memory reset

表 3-4　数控刀具和加工表

		数控刀具和加工表					
材料：铝合金 7075－T6							
刀具号	偏置号	操作	速度 (m/min)	进给 (mm/r)	刀具描述	刀片	TNR
1	01	粗加工： ● 使用 G71 循环指令； ● 粗车切削深度 3.81 mm； ● 留下 0.25 mm 精加工余量	450	0.3	外圆车刀刀杆 MCLNR 12－4	CNMG－432－NR	0.8
2	02	精加工 使用 G70 循环指令	800	0.13	外圆车刀刀杆 MCLNR 12－4	CNMG－434－NF	1.6

表 3-5　指令及其含义

程序模块	指令	含义
—	O2004	
设置工作 数据列表	(..........)	可选的设置数据
	(x0 在主轴中心线上)	
	(z0 在端面上)	
	(..........)	
	(刀具表：)	
	(刀具 1：0.8 TNR 粗加工 TNR)	
	(刀具 2：1.6 精加工)	
	(..........)	
加工启动顺序 换 2 号刀	N0010　G21	米制模式
	N0020　G50 X203.2 Z127	将机床原点设置为①
	N0030　G50 S3000	将主轴最大转速设置为 3 000 r/min
	N0040　T0100	返回位置①换 1 号刀，取消补偿文件 1 中的值
	N0050　G0 M42	将主轴转到中速挡
	N0060　G96 S450M3	主轴 450 m/min 的速度正转旋转，表面恒线速度控制
	(刀具 1：粗加工)	
	N0070　X114.3 Z5 T0101 M8	快速运动到②，读取补偿文件 1 中的值，冷却液开启
G71 之前创建 G42 刀具圆弧半径补偿	N0080　G1 G42 Z2.5 F0.15	按进给量 0.15 运动到③，创建刀具刀尖圆弧半径右补偿

续表

程序模块	指令	含义
操作1： 用G71指令循环 粗加工	N0090 G71 U3.81	开始粗加工 G71 循环,切削深度 3.81 mm
	N0100 G42 G71 P0110 Q0190 U0.6 W0 F0.3	由块 0110～0190 定义的粗糙轮廓。沿 X 轴留下 0.25 mm。进给量 0.3
	N0110 G0 X45.7	快速到④在轮廓上
	N0120 G1 X45.7 z_0	在轮廓上切削到⑤
	N0130 Z－22.9	在轮廓上切削到⑥
	N0140 X66 Z－66	在轮廓上切削到⑦
	N0150 Z－73.7	在轮廓上切削到⑧
	N0160 X78.7	在轮廓上切削到⑨
	N0170 G3 X94z－81.3 r7.6	在轮廓上切削 R3 弧到⑩
	N0180 G1 Z－96.5	在轮廓上切削到⑪
	N0190 X114.3	在轮廓上切削到⑫
	N0200 G0 Z2.5	快速起点③

表 3-6　指令及其含义

—	指令	含义
	（刀具2：精加工）	
G71 指令完成后关闭取消 刀尖圆弧半径补偿	N0210 G1 G40 Z5	进给量下次线性移动时到②,取消刀具刀尖圆弧半径补偿
加工停止	N0220 G0 X203.2 Z127 T0100 M9	快速移动到换刀位置①,取消补偿文件中1的值,关闭冷却液
	N0230 M5	主轴关闭
加工启动顺序 换2号刀	N0240 T0200	换2号刀,取消补偿文件2中的值
	N0250 G0 M43	转换到高挡范围
	N0260 G96 S800 M3	主轴以 800 m/min 的速度正转,恒线速度控制
	N0270 x114.3 z5 T0202 M8	快速到②,读取补偿文件2中的值,冷却液开启
在 G70 循环开始 之前启动 G42	N0280 G1 G42 Z2.5 F0.1	按进给量 0.1 移动到③,并创建刀尖圆弧半径补偿
操作2： 用G70指令循环精加工	N0290 G42 G70 P0110 Q0190	精加工由 0110～0190 定义的轮廓进给量 0.1,数度 800 r/min
G70 循环完成后关闭 G40	N0300 G1 G40 x116	进给量下次线性移动时到12,取消刀尖圆弧半径补偿
加工停止 程序结束	N0310 G0 x203.2 z127 T0200 M9	快速移动到换刀位置①,取消补偿文件中2的值,冷却液关闭
	N0320 M5	主轴关闭
	N0330 M3	程序结束。内存重置

EXAMPLE 3-5

The finished part shown in Figure 3-11 is to be created by boring from stock. Write the CNC program that includes the roughing and finishing cycles. Assume the tool offsets, tool nose radius, and tool nose vector2 have been entered into the offset file. The coordinates of points are list in Table 3-7.

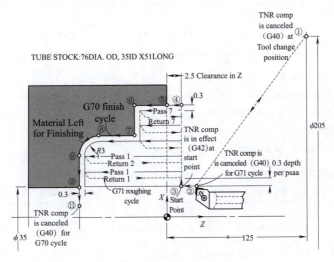

Figure 3-11 Case 3-5

图 3-11 例 3-5

Table 3-7 Points Coordinates

Point	X(mm)	Z(mm)	Point	X(mm)	Z(mm)
②	35	5	⑦	56	−17
③	35	2.5	⑧	56	−33
④	71	2.5	⑨	50	−36
⑤	71	0	⑩	35	−36
⑥	71	−17	⑪	30	−36

■ 例 3-5

加工如图 3-11 所示的零件。编写包含粗加工和精加工循环的数控加工程序。假设刀具刀尖圆弧半径补偿，刀尖圆弧半径和刀尖矢量 2 已输入到补偿文件中。点的坐标列在表 3-7 中。

表 3-7 点的坐标

点	X(mm)	Z(mm)	点	X(mm)	Z(mm)
②	35	5	⑦	56	−17
③	35	2.5	⑧	56	−33
④	71	2.5	⑨	50	−36
⑤	71	0	⑩	35	−36
⑥	71	−17	⑪	30	−36

CNC tool and cutting parameters used to rough turn OD and finish OD contour is shown in Table 3-8. And program to execute the necessary roughing passes and the finish pass is listed in

Table 3-9 and Table 3-10.

粗车外圆和精车外圆的刀具和切削参数见表3-8。加工程序在表3-9和表3-10中。

Table 3-8　CNC Tool and operations sheet

MATERIAL：AL ALLOY 7075 – T6

Tool	Off-Set	Operation	Speed (m/min)	Feed (mm/r)	Tool Description	Insert	TNR
1	01	ROUGH BORE ID ● USE A **G71** ROUGHING CYCLE ● REMOVE 0.3 PER SIDE FOR EACH PASS ● LEAVE0.3 FOR FINISHING	130	0.3	BORING HOLDER S-MCLNR 20 – 4	CNMG – 432 – NR	0.8
2	02	FINISHING ID COUNTOUR ● USE A G71 FINISHING CYCLE	250	0.1	BORING HOLDER MCLNR 20 – 4	CNMG – 434 – NF	1.6

Table 3-9　The Programmings and Meanings

Programming pattern	Word address command	Meaning
—	O2005	Program number
Job setup date listing	(..........)	Optional setup statements
	(X0 IS ON THE SPINDLE CENTERLINE)	
	(Z0 IS ON THE BLANK FACE)	
	(..........)	
	(TOOLING LIST：)	
	(TOOL1：0.8TNR ROUGHING TOOL)	
	(TOOL2：1.6 TNR FINISHING TOOL)	
	(..........)	
Tool2 in turret Machine start-Up sequence	N0010 G21	Metric mode
	N0020 G50 X205 Z125	Set machining origin at①
	N0030 G50 S3000	Set spindle rpm limit to 3 000 r/min
	N0040　T0100	Return to tool change position①. Index to tool 1. Cancel values in offset file 1
	N0050 G0 M42	Shift to intermediate gear range
	N0060 G96 S130 M3	Spindle on（CW）at 130 m/min, constant surface speed
	(TOOL 1：ROUGH TURN CONTOUR)	
	N0070　X35 Z5 T0101 M8	Rapid to②. Coolant on
TNR come on（**G41**）prior to Start of **G71** cycle	N0080 G1 G41 Z2.5F0.3	Ramp on to the left of upward tool motion on the next linear move to③ at feed 0.3

Chapter 3 Techniques and Fixed Cycles for CNC Lathe Programming

续表

Programming pattern	Word address command	Meaning
Operation 1: rough turn contour with G71 cycle	N0090 G71 U0.3	Start rough turning cycle. Cut to 0.5 mm. Per side for each roughing pass
	N0100 G41 G71 P0110 Q0170 U-0.6 W0 F0.3	Rough bore contour defined by blocks 0110 ~ 0170. Leave 0.3 mm Per side along the X axes. Machine at feed 0.3
	N0110 G0 X71	Rapid to ④ on contour
	N0120 G1 X71Z0	Cut to ⑤ on contour
	N0130 Z-17	Cut to ⑥ on contour
	N0140 X56	Cut to ⑦ on contour
	N0150 Z-33	Cut to ⑧ on contour
	N0160 G3 X50 Z-36 R3	Cut R3 arc to ⑨ on contour
	N0170 G1 X35	Cut to ⑩ on contour
	N0180 G0 Z2.5	Rapid to start point ③

Table 3-10 The Programmings and Meanings

—	Word address command	Meaning
	(TOOL2: FINISH TURN CONTOUR)	
TNR comp off (G40) after G71 cycle is completed	N0190 G1 G40 Z2	Ramp off on the next linear move ③ at feed rate
Machine stop squence	N0200 G0 X205Z125T0100 M9	Rapid move to the tool change position ①. cancel values in offset file 1. coolant off
	N0210 M5	Spindle off
Change to tool 2 Machine start-up sequence	N0220 T0200	Index to tool 2. cancel values in offset register 2
	N0230 G0 M43	Shift to high-gear range
	N0240 G96 S250 M3	spindle on (cw) at 250 m/min. Constant surface speed
	N0250 X35 Z5 T0202 M8	Rapid to ②. Coolant on
TNR comp on (G41) prior to start of G70 cycle	N0260 G1 G42 Z1 F0.1	Ramp on to the right of upward tool motion on the next linear move to ③ at feed 0.1
Operation 2: finish bore contour with G70 cycle	N0270 G41 G70 P0 110 Q0 170	Finish bore contour defined by blocks 0110 ~ 0170. Machine at feed 0.1 and speed 250 r/min
TNR comp off (G40) after G70 cycle is completed	N0280 G1 G40 X30	Ramp off on the next linear move to ⑪ at feed rate
Machine stop Program end sequence	N0290 G0 X205Z125T0200 M9	Rapid move to the tool change position ①. Cancel values in offset file 2. coolant off
	N0300 M5	Spindle off
	N0310 M30	Program end. Memory reset

表3-8 数控刀具和加工表

材料:铝合金 7075 – T6

刀具号	补偿号	操作	速度 (m/min)	进给 (mm/r)	刀具描述	刀片	TNR
1	01	粗镗 ● 使用 G71 粗加工循环 ● 每次粗镗切削深度 0.3 ● 留下 0.3 精加工余量	130	0.3	镗刀杆 S-MCLNR 20 – 4	CNMG – 432 – NR	0.8
2	02	精镗 ● 使用 G71 精加工循环	250	0.1	镗刀杆 MCLNR 20 – 4	CNMG – 434 – NF	1.6

表3-9 指令及其含义

程序模块	指令	含义
—	O2005	程序编号
设置工作数据列表	(..........) (X0 在主轴中心线上) (Z0 在端面上) (..........) (刀具表:) (刀具 1:0.8 TNR 粗加工刀具) (刀具 2:1.6 TNR 精加工刀具) (..........)	可选的设置数据
加工启动顺序 换 2 号刀	N0010 G21	米制模式
	N0020 G50 X205 Z125	将机床原点设置为①
	N0030 G50 S3000	将主轴最大转速设置为 3 000 r/min
	N0040 T0100	在位置①换 1 号刀,取消补偿文件 1 中的值
	N0050 G0 M42	将主轴转到中速挡
	N0060 G96 S130 M3	主轴以 130 m/min 速度正转
	(刀具 1:粗加工)	
	N0070 X35 Z5 T0101 M8	快速到②,读取补偿文件 1 中的值,冷却液开启
开始 G71 循环之前启动 G41	N0080 G1 G41 z2.5 F0.30	创造刀具圆弧半径左补偿,进给量 0.3 刀具运动到③

续表

程序模块	指令	含义
操作1: 用G71指令 循环粗加工	N0090 G71 u0.3	开始粗加工 G71 循环,每次切削深度 0.3 mm
	N0100 G41 G71 P0110 Q0170 U−0.6w0 F0.3	粗车轮廓由 0110～0170 定义。沿 X 轴留下 0.3 mm 精车余量,机床进给量 0.3 mm/r
	N0110 G0 X71	快速到④在轮廓上
	N0120 G1 X71 Z0	在轮廓上切削到⑤
	N0130 Z−17	在轮廓上切削到⑥
	N0140 X56	在轮廓上切削到⑦
	N0150 Z−33	在轮廓上切削到⑧
	N0160 G3 x50 Z−36 R3	在轮廓上切削 R3 弧到⑨
	N0170 G1 X35	在轮廓上切削到⑩
	N0180 G0 Z2.5	快速回到起点③

表 3-10　程序及其含义

	指令	含义
—	(刀具2:精加工)	
在 G71 完成后 运行 G40 指令	N0190 G1 G40 Z2	以进给量在下一个线性移动到③取消刀补
加工停止	N0200 G0 X205 Z125 T0100 M9	快速移动到换刀位置①,取消补偿文件 1 中的值,冷却液关闭
	N0210 M5	主轴关闭
加工启动顺序 换2号刀	N0220 T0200	换 2 号刀,取消补偿文件 2 中的值
	N0230 G0 M43	转向高档范围
	N0240 G96 S250 M3	主轴以 250 m/min 速度正转
	N0250 X35 Z5 T0202 M8	快速到②,读取补偿文件 2 中的值,冷却液开启
在开始 G70 循环之前 启动 G41	N0260 G1 G42 Z2.5 F0.1	进给量 0.1,在下一次线性移动到③时,完成刀尖圆弧半径补偿
操作2:用 G70 指令精加工 孔轮廓	N0270 G41 G70 P0 110 Q0 170	完成由 0110～0170 定义的孔轮廓,进给量 0.1 mm/r,转速 250 m/min
G70 循环完成之后 关闭 G40	N0280 G1 G40 X30	下一次线性移动到⑪,取消刀尖圆弧半径补偿
加工停止, 程序结束	N0290 G0 X205 Z125 T0200 M9	快速移动到换刀位置①,取消补偿文件 2 中的值,冷却液关闭
	N0300 M5	主轴关闭
	N0310 M30	程序结束。内存重置

3.3.3 Peck Drilling and Face Grooving Cycle：G74

3.3.3 钻孔和端面切槽循环指令：G74

The G74 word initiates a multipurpose cycle that can be used for peck drilling the stock center as well as peck machining wide or multiple face grooves. It is a simple cycle that does not produce a high-quality machine finish. Applications include drilling a hole in solid stock prior to ID boring or threading operations and for roughing deep grooves prior to finish machining.

G74 是用于钻削中心孔以及端面切槽的复合循环指令。这个循环不能产生高质量的表面粗糙度。该指令适用于钻孔或螺纹加工之前在实心毛坯上钻底孔，以及在精加工之前的粗加工切槽。

1. Peck Drilling the Part Center

1. 钻削零件中心孔

 **General Syntax**

G74 Rn（block1）

G74 X0 Zn Qn R0 Fn（Block2）

Rn The numeric value of n specifies the rapid retract amount after each peck.

G74 Initiates the peck drilling cycle when programmed with the address：X0 Zn Qn R0 Fn, See Figure 3-12.

X0 is the center of the part along the X-axis.

Zn The numeric value of n specifies the final depth of the drilled hole along the axis.

Qn The numeric value of n specifies the first peck distance below the clearance plane. This value is added successively to the last total for each pass until the final hole depth is reached.

R0 Address specifies relief amount at the end of each cut. Value must be set to 0.

Fn The numeric value of n specifies the feed rate of the tool.

常用格式

G74 Rn（block1）

G74 X0 Zn Qn R0 Fn（Block2）

Rn n 的数值是指每次切削后的返回量。

G74 使用以下地址编程时启动钻削循环：X0、Zn、Qn、R0 和 Fn，如图 3-12 所示。

X0 是沿 X 轴的零件中心。

Zn n 的数值指定钻孔的最终深度。

Qn n 的数值指每次 Z 向切削深度，使用此数值沿 Z 向进行多次切削直到获得所需孔深。

R0 指每次切削结束时的退刀量，值必须设置为 0。

Fn n 指刀具的进给量。

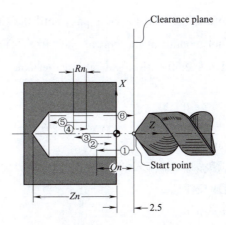

Figure 3-12　The Cutting Path of G74
图 3-12　G74 的切削路径

NOTES：

（1）The G74 peck drilling does not retract the tool out of the hole after each peck as does the G83 peck cycle. It can only break chips between pecks and not clear them out of the hole. The G74 cycle is recommended for holes that are not too deep. Depths approximately less than four times the drill diameter.

（2）Constant spindle RPM must be sat by G97 Sn block coded prior to programming the G74 peck drilling cycle.

（3）The drilling tool must be positioned the start point on the clearance plane prior to initiating the G74 peck drilling cycle. The X-axis coordinate is X0 at this point.

注意事项：

（1）与 G83 钻孔循环一样，G74 钻孔不会在每次钻孔后将刀具从孔中退出。只能将切屑打断而不会将它们从孔中清除。对于不太深的孔，当深度大约小于钻头直径的 4 倍时，建议使用 G74 循环。

（2）在编程 G74 钻孔循环之前，必须通过 G97 设置恒主轴转速。

（3）在开始 G74 钻孔循环之前，必须将钻孔刀具定位在起始平面，此时 X 轴坐标为 X0。

2. Wide Face Grooving
2. 车端面宽槽和多沟槽

 General Syntax

G74 Rn（BLOCK1）

G74 Xn Zn Pn Qn R0 Fn（BLOCK2）

G74 Initiates the grooving cycle when programmed with the addresses：Xn Zn Pn Qn Rn Fn.

Rn　The numeric value of n specifies the return amount after each peck。

Zn　The numeric value of n specifies the final depth of the groove along the Z-axis.

Qn　The numeric value of n specifies the first peck distance below the clearance plane. This value is added successively to the last total for each pass until the final groove depth is reached.

R0 Address specifies relief amount at the end of each peck. Value must be set to 0.

Fn The numeric value of n specifies the feed rate of the tool.

常用格式

G74 Rn（BLOCK1）

G74 Xn Zn Pn Qn R0 Fn（BLOCK2）

G74 使用以下地址启动切槽循环：Xn、Zn、Pn、Qn、Rn 和 Fn。

Rn n 的数值指每次切削后的返回量。

Zn n 指定沿 Z 轴的槽的最终深度。

Qn n 的数值指每次切槽的深度，该值被应用到每次切削，直到达到最终槽深。

R0 指定每次切槽结束时的退刀量，值必须设置为 0。

Fn n 指定刀具的进给量。

（1）Wide Grooves

（1）宽槽

Xn The numeric value of n specifies the final X-position of the tool for the cycle where n is in terms of diameter. See in Figure 3-13.

Pn The numeric value of n specifies the step over distance on the X-axis. The tool is successively offset this distance each time it reaches the groove bottom and is rapided to the clearance plane. See Figure 3-13.

Xn n 指刀具的切槽 X 向最终位置，其中 n 是直径值，如图 3-13 所示。

Pn n 指 X 轴上的步距。刀具连续地偏移这个距离，切削到达凹槽底，然后回到起始平面，如图 3-13 所示。

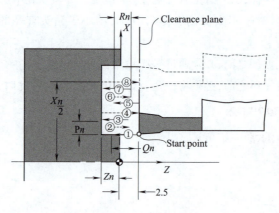

Figure 3-13　The Cutting Path of G74

图 3-13　G74 的切削路径

（2）Multiple Grooves

（2）多沟槽

Xn The numeric value of n specifies the location of the last groove on the X-axis where n is in terms of diameter. See Figure 3-14.

Pn The numeric value of **n** specifies the distance between grooves on the *X*-axis. See Figure 3-14.

Xn n 指定 X 轴上最后一个凹槽的位置,为直径,如图 3-14 所示。

Pn n 指定 X 轴上的凹槽之间的距离,如图 3-14 所示。

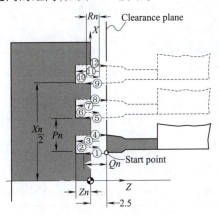

Figure 3-14　The Cutting Path of G74
图 3-14　G74 的切削路径

NOTES:

(1) The tool control point influences the value programmed for Xn. In these illustrations it is assumed to be at the lower left corner, see in Figure 3-15. The programmer must determine which corner of the tool is the leading edge control point.

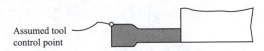

Figure 3-15　Tool Control Point
图 3-15　刀位点

(2) The control uses the tool positioning block, which appears just before the G74 block, to establish the proper start point for machining the groove (s).

注意事项:

(1) 刀位点影响 Xn 编程值。在图 3-15 中,假设它位于左下角。程序员必须确定该刀具的哪个角是刀位点。

(2) 控制器使用出现在 G74 程序段之前的刀具定位程序段,以确定加工凹槽的起始点。

■ **EXAMPLE 3-6**

Write a word address program for peck drilling the center of the part mounted on a turning center. See Figure 3-16.

■ **例 3-6**

写一段程序,用于加工安装在车削中心上的零件的中心孔,如图 3-16 所示。

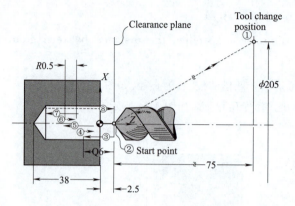

Figure 3-16 Case 3-6

图 3-16 例 3-6

The program to execute the necessary peck drilling passes is listed in Table 3-11.

加工孔的啄钻指令列在表 3-11 中。

Table 3-11 The Programmings and Meanings

Programming pattern	Word address command	Meaning
	O2006	Program number
	(..........)	Optional setup statement
Tool 1 in turret Machine start-Up sequence	N0010 G21	Metric mode
	N0020 G50 X205 Z75	Set machining origin at ①
	(TOOL1 :12 DIA LATHE DRILL)	Optional comment statement
	N0030 T0100	Return to tool change position ①. Index to tool 1. Cancel values in offset register 1
	N0040 G0 M41	Shift to low gear range
	N0050 G97 S400 M3	spindle on (cw) at 400 r/min. Constant surface speed canceled
	N0060 X0 Z2.5 T0101 M8	Rapid tool to start point ②. use offset values in register 1. coolant on
Operation 1: Peck drill the center of the stock	N0070 G74 R0.5	Initiate G74 cycle. Set the rapid retract distance 0.5 after each cut
	N0080 G74 X0 Z-38 Q6000 R0 F0.3	Peck drill to a depth of 38mm. at a feed rate of 0.3. Set the peck depth at 6 mm
Machine stop Program end sequence	N0090 G0 X205 Z77.5 T0100 M9	Cancel cycle. Rapid to tool change position ①. Cancel values in offset register 1
	N0100 M5	Spindle off
	N0100 M30	Program end. Memory reset

Chapter 3 Techniques and Fixed Cycles for CNC Lathe Programming

表 3-11 指令及其含义

程序模块	指令	含义
	O2006	程序编号
	(..........)	可选的设置数据
加工启动顺序 换1号刀	N0010 G21	公制模式
	N0020 G50 X205 Z75	将机床原点设置为①
	（工具:φ12 的车床用钻头）	可选注释语句
	N0030 T0100	返回换刀位置①,换 1 号刀,取消补偿文件 1 中的值
	N0040 G0 M41	主轴切换到低速挡
	N0050 G97 S400 M3	主轴 400 r/min 速度正转
	N0060 X0 Z2.5T0101 M8	快速移动刀具到②,读取补偿文件 1 中的值,冷却液打开
操作 1: 钻中心孔	N0070 G74 R0.5	启动 G74 循环,每次切削后快速回退 0.5 mm
	N0080 G74 X0 Z-38Q6000 R0 F0.3	钻孔深度 38 mm.进给量为 3 mm/r.将每次切削深度设置为 6 mm
加工停止, 程序结束	N0090 G0 X205 Z75T0100 M9	取消循环.快速运动至换刀位置①。取消补偿文件 1 中的值
	N0100 M5	主轴关闭
	N0100 M30	程序结束。内存重置

■ **EXAMPLE 3-7**

Write a word address program to machine a wide groove into the face of the part mounted on a turning center, as shown in Figure 3-17.

■ **例 3-7**

写一个程序,将安装在车削中心的零件上加工一个宽槽,如图 3-17 所示。

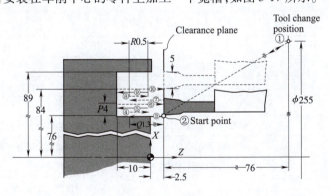

Figure 3-17 Case 3-7
图 3-17 例 3-7

The program to execute the necessary grooving passes is listed in Table 3-12.
加工槽的指令列在表 3-12 中。

Table 3-12 The Programmings and Meanings

Programming pattern	Word address command	Meaning
Tool 1 in turret Machine start-Up sequence	O2007	Program number
	(..........)	Optional setup statement
	N0010 G21	Metric mode
	N0020 G50 X255 Z76	Set machining origin at①
	N0030 G50 S3000	Set spindle rpm limit to 3 000 r/min
	(TOOL3：WIDE GROOVE AS PERPRINT)	Optional comment statement
	N0040 T0300	Return to tool change position①. Index to tool 3. Cancel values in offset register1
	N0050 G0 M42	Shift to intermediate gear range
	N0060 G96 S215 M3	spindle on (cw) at 215 m/min. Constant surface speed
	N0070 X152 Z2.5 T0303 M8	Rapid tool to start point②. use offset values in register3. coolant on
Operation 1： Peck wide face groove into stock.	N0070 G74 R0.5	Initiate G74 cycle. Set the rapid retract distance 0.5 after each cut
	N0080 G74 X168 Z-10P4000 Q1300 R0 F0.25	Peck wide grooves starting at the smaller diameter of 152 mm. Final position of the tool is at 168mm. Set groove depth to 10mm and stepover distance to 4mm. Set the peck depth to 1.3mm at feed rate 0.25 mm/r
Machine stop Program end sequence	N0090 G0 X255Z76T0300 M09	Cancel cycle. Rapid to tool change position①. Cancel values in offset register 3
	N0100 M5	Spindle off
	N0110 M30	Program end. Memory reset

表3-12 指令及其含义

程序模块	指令	含义
加工启动顺序 换1号刀	O2007	程序编号
	(..........)	可选的设置数据
	N0010 G21	公制模式
	N0020 G50 X255 Z76	将机床原点设置为①
	N0030 G50 S3000	将主轴最大转速设置为3 000 r/min
	(刀具3：每个切削的宽度)	可选注释语句
	N0040 T0300	返回换刀位置①，换3号刀，取消补偿文件1中的值
	N0050 G0 M42	将主轴转到中速挡
	N0060 G96 S215 M3	主轴以215 m/min 速度正转
	N0070 X152 Z2.5 T0303 M8	快速移动刀具到②，使用补偿文件3中的值，冷却液打开

续表

编程模式	指令	含义
操作1：加工端面凹槽	N0070 G74 R0.5	启动 G74 循环，每次切削后快速回退 0.5 mm。
	N0080 G74 X168 Z−10 P4000 Q1300 R0 F0.25	从 152 mm 的较小直径开始啄宽凹槽，刀具的最终位置为 168 mm，将凹槽深度设置为 10 mm，将步距设置为 4 mm。将每次切削深度设置为 1.3 mm，进给量 0.25 mm/r
加工停止，程序结束	N0090 G0 X255 Z76 T0300 M09	取消循环。快速运动到换刀位置①。取消补偿文件 3 中的值
	N0100 M5	主轴关闭
	N0110 M30	程序结束。内存重置

The machining can also be started at the larger diameter of 168 mm. In this case the startup blocks would be：

加工也可以在 168 mm 的较大直径下开始。在这种情况下，启动程序段为：

N0060 G0 X168 Z2.5 T0303 M8

N0070 G74 R0.5

N0080 G74 X152 Z−10 P4000 Q1300 R0 F0.25

■ **EXAMPLE 3-8**

For the part given in Example 3-8, instead of a single wide groove, machine a series of grooves into the end face, as shown in Figure 3-18.

■ **例 3-8**

对于例 3-8 中给出的零件，要在端面上加工多个凹槽而不是单个宽槽，如图 3-18 所示。

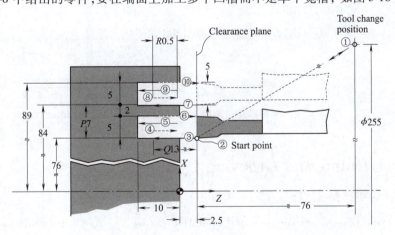

Figure 3-18　Case 3-8

图 3-18　例 3-8

The same word address program applies as given in Example 3-8 except that the G74 block listed in Table 3-13：

除了表 3-13 中的 G74 程序段，其他程序段与例 3-8 相同。

Table 3-13　The Programmings and Meanings

	Word address command	Meaning
Operation 1: Peck a series of face groove into Stock.	N0070 G74 R0.5	Initiate G74 cycle. Set the rapid retract distance 0.5 mm after each cut
	N0080 G74 X168 Z-10 P7000 Q1300 R0 F0.25	Peck a series of face grooves starting at the smaller diameter of 152 mm. Final X position of the tool is at 168 mm. Set groove depth to 10 mm and distance between each groove to 7 mm. Set peck depth to 1.3 mm at feed rate 0.25 mm/r

表 3-13　指令及其含义

	指令	含义
操作 1: 在端面上加工槽	N0070 G74 R0.5	启动 G74 循环。每次切削后快速回退 0.5 mm
	N0080 G74 X168Z-10 P7000Q1300 R0 F0.25	从 152 mm 的较小直径开始加工出一系列面槽。刀具的最终 X 位置为 168 mm。将槽深设置为 10 mm，每个槽之间的距离为 7 mm。将切削深度设置为 1.3 mm，进给量是 0.25 mm/r

If the tool control point were located at the upper-left corner of the tool, as shown in Figure 3-19, and the machining started at the larger diameter of 178 mm, the startup blocks would be：

如果刀位点位于刀具的左上角，如图 3-19 所示，并且加工从直径较大的 178 mm 开始，则启动程序段为：

N0060 G0 X178 Z2.5T0303 M8
N0070 G74 R0.5
N0080 G74 X162 Z-10 P7000Q1300 R0 F0.25

G75切槽或切断循环指令（切槽）2D轨迹

Figure 3-19　Assumed tool control point

图 3-19　假设的刀位点

3.3.4　Peck Cutoff and Grooving Cycle：G75

3.3.4　切断和切槽循环指令：G75

The G75 cycle is used to execute cutoff operations as well as rough machining wide and multiple grooves. It is identical to the G74 cycle if the *X* and *Z* axes and the *P* and *Q* addresses were reversed.

G75 循环用于执行切断操作以及粗加工多个凹槽。如果 *X* 轴和 *Z* 轴以及 *P* 和 *Q* 地址对调，则与 G74 循环相同。

Chapter 3 Techniques and Fixed Cycles for CNC Lathe Programming

1. Cutoff

1. 切断

 **General Syntax**

G75 Rn（BLOCK1）

G75Xn Zn Pn R0 Fn（BLOCK2）

G75 Initiates the cutoff cycle when programmed with the addresses：Xn Zn Pn R0 Fn. See Figure 3-20.

Rn The numeric value of n specifies the rapid retract amount after each peck

Xn The numeric value of n specifies the distance the tool moves below the center of the part along the X-axis when executing the cutoff groove. Here, n is a negative value expressed in terms of diameter.

Zn The numeric value of n specifies the location of the cutoff along the Z-axis.

Pn The numeric value of n specifies the first peck distance below the clearance plane. This value is added successively to the last total for each pass until the final groove depth is reached.

R0 Address specifies relief amount at the end of each cut. Value must be set to 0.

Fn The numeric value of n specifies the feed rate of the tool.

G75切槽
或切断循环
指令（切断）
2D轨迹

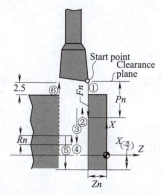

G75车槽固定
循环指令
3D视频

Figure 3-20　The Cutting Path of G75

图 3-20　G75 的切削路径

常用格式

G75 Rn（BLOCK1）

G75 Xn Zn Pn R0 Fn（BLOCK2）

G75 启动切断循环：Xn、Zn、Pn、R0 和 Fn，如图 3-20 所示。

Rn n 指每次切削后的退刀量。

Xn 数值 n 指当执行切断时，刀具沿 X 轴在零件中心以下移动的距离，n 是以直径表示的负值。

Zn n 的数值指沿 Z 轴切断的位置。

Pn n 的数值指每次切削深度，使用该值连续切削直到最终切断深度。

R0 地址指定每次切削结束时的退刀量,该值必须设置为 0。

Fn n 指定刀具的进给量。

2. Wide and Multiple Grooving

2. 宽槽和多沟槽

General Syntax

G75 Rn (BLOCK1)

G75Xn Zn Pn Qn R0 Fn (BLOCK2)

G75 Initiates the grooving cycle when programmed with the addresses: Xn Zn Pn Qn R0 Fn.

Rn The numeric value of n specifies the rapid retract amount after each peck.

Xn The numeric value of n specifies the final depth of the groove along the X-axis, where n is in terms of diameter.

Pn The numeric value of n specifies the first peck distance below the clearance plane. This value is added successively to the last total for each pass until the final groove depth is reached.

R0 Address specifies relief amount at the end of each peck. Value must be set to 0.

Fn The numeric value of n specifies the feed rate of the tool.

常用格式

G75 Rn (BLOCK1)

G75Xn Zn Pn Qn R0 Fn (BLOCK2)

G75 使用以下地址编程时启动切槽循环:Xn、Zn、Pn 、Qn、R0 和 Fn。

Rn n 指每次切槽后的退刀量。

Xn n 指定沿 X 轴的凹槽的最终深度,其中 n 是直径。

Pn n 是每次切削深度,使用该值连续切削直到最终深度。

R0 指每次切槽结束时的退刀量,该值必须设置为 0。

Fn n 指刀具的进给量。

(1) Wide Grooves

(1) 宽槽

Zn The numeric value of n specifies the final Z-position of the tool for the cycle. See in Figure 3-21.

Qn The numeric value of n specifies the step over distance on the Z-axis. The tool is successively offset this distance each time it reaches the groove bottom and is rapided to the clearance position. See Figure 3-21.

Zn n 指槽的最终 Z 向位置,如图 3-21 所示。

Qn n 指 Z 轴上的距离。刀具连续地偏移这个距离,切到凹槽底部,并返回起始平面,如图 3-21 所示。

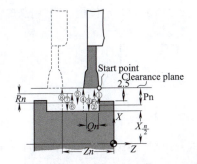

Figure 3-21　The Cutting Path of G75

图 3-21　G75 的切削路径

（2）Multiple Grooves

（2）多沟槽

Zn　The numeric value of n specifies the location of the last groove on the Z-axis. See in Figure 3-22.

Qn　The numeric value of n specifies the distance between grooves on the Z-axis. See Figure 3-22.

Zn　n 指 Z 轴上最后一个凹槽的位置，如图 3-22 所示。

Qn　n 指 Z 轴上凹槽之间的距离，如图 3-22 所示。

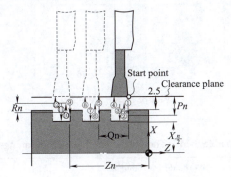

Figure 3-22　The Cutting Path of G75

图 3-22　G75 的切削路径

Notes：

（1）The tool control point influences the value programmed for Zn. In this illustration it is assumed to be at the lower right corner(Figure 3-23). The programmer must determine which corner of the tool is the leading edge control point.

（2）The controller uses the tool positioning block which appears just before the G75 block to establish the proper start point for machining the cutoff or groove(s).

注意事项：

（1）刀位点影响 Zn 的数值。在图 3-23 中，假设它位于右下角。程序员必须确定刀具的哪个角是刀位点。

(2) 控制器在 G75 程序段运行之前的刀具定位程序段,用来确定切断或切削凹槽的起始点。

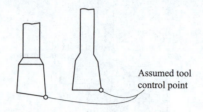

Figure 3-23　Assumed Tool Control Point

图 3-23　假设的刀位点

■ EXAMPLE 3-9

Write a word address program in table 3-14 to execute peck cutoff of the part shown in Figure 3-24.

■ 例 3-9

在表 3-14 中写一段程序来执行图 3-24 所示部件的切断操作。

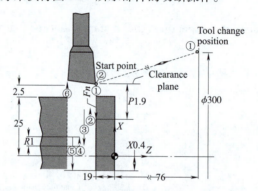

Figure 3-24　Case 3-9

图 3-24　例 3-9

Table 3-14　The Programmings and Meanings

Programming pattern	Word address command	Meaning
Tool 6 in turret Machine start- Up sequence	O2009	Program number
	(..........)	Optional setup statement
	N0010 G21	Metric mode
	N0020 G50 X300 Z76	Set machining origin at①
	N0030 G50 S3000	Set spindle rpm limit to 3 000 r/min
	(TOOL6: CUTOFF AS PER PRINT)	Optional comment statement
	N0040 T0600	Return to tool change position①. Index to tool 6. Cancel values in offset register 6
	N0050 G0 M42	Shift to intermediate gear range
	N0060 G96 S100 M3	spindle on (cw) at 200 m/min. Constant surface speed
	N0070 G0 X55 Z-19 T0606 M8	Rapid tool to start point②. use offset values in register6. coolant on

续表

Programming pattern	Word address command	Meaning
Operation 1: Peck cutoff the Stock.	N0080 G75 R1	Initial G75 cycle. Set the rapid retract distance 1 mm after each cut
	N0090 G75 X−0.8P1900 R0 F0.1	Peck cutoff at location Z−19mm at pect feed rate of 1 mm/r. Set the peck depth at 1.9 mm. set the peck feed rate to 0.1 mm/r.
Machine stop Program end sequence	N0100 G0 X300 Z76 T0600 M9	Cancel cycle. Rapid to tool change position ①. Cancel values in offset register 1
	N0110 M5	Spindle off
	N0120 M30	Program end. Memory reset

表 3-14 指令及其含义

程序模块	指令	含义
	O2009	程序编号
	(..........)	可选的设置数据
加工启动顺序 换 6 号刀	N0010 G20	公制模式
	N0020 G50 X300 Z76	将机床原点设置为①
	N0030 G50 S3000	将主轴转速限制设置为 3 000 r/min
	(刀具6:每个切削的宽度)	可选注释语句
	N0040 T0600	返回换刀位置①,换6号刀,取消补偿文件6中的值
	N0050 G0 M42	将主轴转到中速挡
	N0060 G96 S100 M3	主轴 200 m/min 速度正转
	N0070 G0 X55 Z−19 T0606 M8	快速移动到起始点②。使用补偿文件 6 中的值,冷却液打开
操作 1:切断	N0080 G75 R1	启动 G75 循环,每次切削后快速回退 1 mm
	N0090 G75 X−0.8 P1900R0 F0.1	切断位置为 Z−19 mm,将切削深度设置为 1.9 mm,切削进给量设置为 0.1 mm/r
加工停止 程序结束	N0100 G0 X300 Z76 T0600 M9	取消循环。快速运动到换刀位置①。取消补偿文件 1 中的值
	N0110 M5	主轴关闭
	N0120 M30	程序结束。内存重置

■ **EXAMPLE 3-10**

A wide groove is to be peck machined into the part shown in Figure 3-25. Write the required word address program in Table 3-15.

■ **例 3-10**

在图 3-25 所示的零件上加工一个宽槽,在表 3-15 中写下所需的程序。

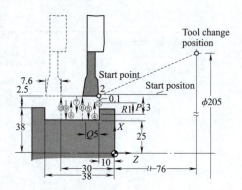

Figure 3-25　Case 3-10

图 3-25　例 3-10

Table 3-15　The Programmings and Meanings

Programming pattern	Word address command	Meaning
	O2010	Program number
	(..........)	Optional setup statement
Tool 6 in turret Machine start-up sequence	N0010 G21	Metric mode
	N0020 G50 X205 Z76	Set machining origin at ①
	N0030 G50 S3000	Set spindle rpm limit to 3 000 r/min
	(TOOL5：WIDE GROOVE AS PER PRINT)	Optional comment statement
	N0040 T0500	Return to tool change position ①. Index to tool 5. Cancel values in offset register 5
	N0050 G0 S100 M42	Shift to intermediate gear range
	N0060 G96 S100 M3	spindle on (cw) at 100 m/min. Constant surface speed
	N0070 G0 X81 Z−10T0505 M8	Rapid tool to start point ②. use offset values in register6. coolant on
Operation 1: Peck cutoff the Stock.	N0080 G75 R1	Initiate G75 cycle. Set the rapid retract distance 1mm after each cut
	N0090 G75 X50 Z−30P1300 Q5000 R0 F0.1	Peck wide groove to 50 mm in diameter at feed rate 0.1 mm/r. Final Z position of the tool is at −30 mm. The peck depth is 1.3 mm in stepover distance is set to 5 mm
Machine stop Program end Sequence.	N0100 G0 X205 Z76 T0500 M9	Cancel cycle. Rapid to tool change position ①. cancel values in offset register 1
	N0110 M5	Spindle off
	N0120 M30	Program end. Memory reset

表 3-15 指令及其含义

程序模块	指令	含义
	O2010	程序编号
	(..........)	可选的设置数据
加工启动顺序 换 6 号刀	N0010 G20	米制模式
	N0020 G50 X203 Z76	将机床原点设置为①
	N0030 G50 S3000	将主轴转速限制设置为 3 000 r/min
	(刀具 5：每个切削的宽度)	可选注释语句
	N0040 T0500	返回换刀位置①，换 5 刀，取消补偿文件 5 中的值
	N0050 G0 S100 M42	将主轴转到中速挡
	N0060 G96 S100 M3	主轴为 100 m/min 速度正转
	N0070 G0 X81 Z-10 T0505 M8	快速运动到②，读取补偿文件 6 中的值，冷却液打开
操作 1：切槽	N0080 G75 R1	发起 G75 循环，每次切削后快速回退 1 mm
	N0090 G75 X50 Z-30 P1300 Q5000 R0 F0.1	在进给量 0.1 下切槽宽至直径 50 mm，刀具最终 Z 位置为 -30 mm。切削深度为 1.3 mm，侧向步距为 5 mm
加工停止， 程序结束	N0100 G0 X205 Z76 T0500 M9	取消循环，快速回换刀位置①，取消补偿文件 1 中的值
	N0110 M5	主轴关闭
	N0120 M30	程序结束。内存重置

■ **EXAMPLE 3-11**

Instead of machining a single wide groove into the part given in Example 3-10, produce a series of grooves as shown in Figure 3-26.

■ 例 3-11

如图 3-26 所示，用多槽代替在示例 3-10 中给出单个宽槽进行加工。

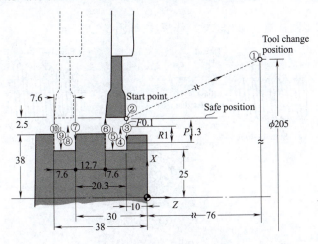

Figure 3-26　Case 3-11

图 3-26　例 3-11

The same word address program applies as given in Example 3-10 except that the G74 block listed in Table 3-16.

除了 G74 指令，其他程度指令与例 3-10 相同。G74 指令列在表 3-16 中。

Table 3-16 The Programmings and Meanings

Programming Pattern	Word address command	Meaning
Operation1: peck a series of grooves into stock.	N0070 G75 R1	Initiate G74 cycle. Set the rapid retract distance 1mm after each cut
	N0080 G75 X50 Z−30 P1300 Q20300 R0 F0.1	Pect a series of face grooves to 50 mm in diameter at feed rate of 0.1 mm/r. Final Z position of the tool is at −30 mm。The peck depth is set to 1.3 mm. Distance between grooves is set to 20.3 mm

表 3-16 指令及其含义

程序模块	指令	含义
操作 1：切槽	N0070 G75 R1	启动 G74 循环。每次切削快速退回量 1 mm
	N0080 G75 X50 Z−30 P1300 Q20300 R0 F0.1	槽直径为 50 mm，进给量速率为 0.1 mm/r，刀具最终 Z 位置为 −30 mm，切削深度设置为 1.3 mm，凹槽之间的距离设置为 20.3 mm

3.4　Thread Cutting on CNC Lathes and Turning Centers

3.4　数控车床及车削中心中的螺纹加工

Turning centers can be programmed to machine straight, tapered, or scroll threads. Refer to Figure 3-27.

车削中心可以加工圆柱螺纹、圆锥螺纹、线螺纹，如图 3-27 所示。

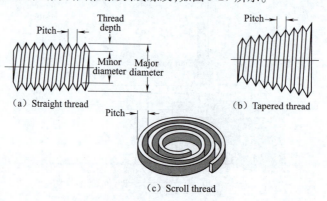

Figure 3-27　Machine Straight, Tapered, or Scroll threads

图 3-27　加工圆柱螺纹、圆锥螺纹、线螺纹

Threads are machined by a special tool possessing the thread shape. The tool is positioned at a specific Z starting distance from the end of the work, a distance that varies from machine to machine. Its value can be found in the machine's programming manual. A good rule of thumb to follow is to use a starting distance of approximately four times the thread pitch. Beginning at the start point, the servo motors will accelerate the tool to a feed rate equal the required thread lead. The tool creates the thread shape by repeatedly following the same path as axial infeed is applied. For standard V threads the infeed can be applied along a 0°, 29°, or 30° angle. The depth of cut for the first pass is largest, and is decreased for each successive pass until the required thread depth is achieved. Decreasing the depth per pass insures the sensitive thread profile point of the cutting edge is not overloaded as the insert cuts deeper to form more and more of the profile.

螺纹由具有螺纹形状的特殊刀具加工而成。该刀具位于距工件端面特定的 Z 轴起始距离处，该距离因机床而异。这个距离值可以在机床的编程手册中找到。根据经验可以使用约等于螺距的四倍的起始距离。从起始点开始，电机将刀具加速到与所需螺纹导程相等的进给量。刀具通过重复切削来加工出螺纹。对于标准 V 螺纹，可以沿 0°、29° 或 30° 角进给。第一次的切削深度最大，并且每次连续切削深度逐渐减少，直到达到所需的螺纹深度。减小每次切削工序的深度，可以确保切削刃的螺纹不会过切，因为刀片切得越深，切削力越大。

3.4.1 Single-Pass Threading Cycle：G32

3.4.1 单线螺纹循环指令：G32

The most basic threading cycle is initiated by a G32 word. Precise control over the depth of each thread pass, as well as the infeed angle, are afforded. The cycle requires four blocks of data to perform one threading pass (Figure 3-28), and the tool must be positioned at the appropriate start point prior to executing the cycle.

最基本的螺纹加工指令是 G32，能精确控制每次螺纹切削深度以及进给角度。循环需要四个程序段来执行一次螺纹的加工（图 3-28），并且必须在执行循环之前将刀具定位在适当的起始点。

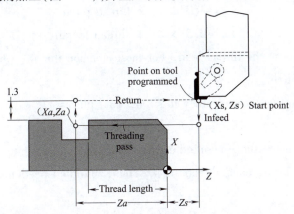

Figure 3-28 The Cutting Path of Maching a Thread Pass

图 3-28 加工一次螺纹线程的切削路径

General Syntax

G0 Xa

G32 Za Fn

G0 Xs

Zs

G32 Initiates the single-pass threading cycle.

Xa Specifies the absolute X coordinate of the tool after axial infeed.

Za Specifies the absolute Z coordinate of the tool after the threading pass.

Fn The value of n specifies the feed rate of the tool.

Xs Zs Specifies the absolute X and Z coordinates of the start point.

常用格式

G0 Xa

G32 Za Fn

G0 Xs

Zs

G32 启动单线螺纹循环。

Xa 指轴向进刀后刀具的 X 轴绝对坐标值。

Za 指螺纹加工时刀具的 Z 轴绝对坐标值。

Fn n 指刀具的进给量。

Xs Zs 指起始点 X 轴和 Z 轴的绝对坐标值。

NOTES：

（1）The starting and stopping distance Zs, Za can be determined from the manufacturer's programming manual or estimated by the formulas：

$$Zs \approx (3 \text{ to } 4) \times \text{thread pitch}$$

$$Za \approx 0.5 \times Zs + \text{thread length}$$

（2）Fn must be set to the thread lead. For most common thread, which is single, start the lead the same as the pitch.

The value of n can have an accuracy to six decimal places.

$$n = \text{Lead} = \text{pitch}$$

（3）In order to ensure that the spindle speed and feed rate operate in proper unison, both must remain constant during each thread pass. Thus, spindle speed must be set with a G97（r/min）word.

（4）Manual adjustment of the feed rate is not possible when a G32 cycle is in effect.

（5）Straight tapered, and scroll threads may be cut. The cycle can be used for outside as well as inside threading.

(6) The infeed direction is controlled by the programmer.

注意事项:

(1) 起始距离 Zs 和终止距离 Za,可以从制造商的编程手册中确定或由公式估算:

$$Zs \approx (3 \sim 4) \times 螺距$$

$$Za \approx 0.5 \times Zs + 螺纹长度$$

(2) 必须将 Fn 设置为螺纹导程。对于大多数常见的螺纹是单线螺纹,它的导程与螺距一致。

$$n = 导程 = 螺距$$

n 的值可以精确到小数点后六位。

(3) 为了确保主轴转速和进给量保持一致地工作,在每个螺纹车削期间,两者必须保持恒定。因此,必须使用 G97(r/min) 设置主轴转速。

(4) 当 G32 循环生效时,无法手动调节进给量。

(5) 可以切削锥螺纹和线螺纹。该循环可用于切削外螺纹和内螺纹。

(6) 进给方向由程序员控制。

■ EXAMPLE 3-12

Threads are to be cut into the part blank, as indicated in Figure 3-29. Use the G32 cycle for writing the program segment.

■ 例 3-12

在工件上切削螺纹,如图 3-29 所示,使用 G32 循环编写程序段。

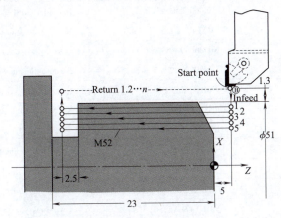

Figure 3-29　Case 3-12

图 3-29　例 3-12

The programmer has elected to use five passes for machining the threads to depth. The infeed per pass can be found from the formula:

$$\text{Infeed}_p = \frac{d}{\sqrt{p-1}} \times \sqrt{j}$$

where, infeed$_p$ is radial feed, d is the depth of the threads, p is the number of passes, $j = 0.3$ for first pass, $j = 1$ for second pass and $j = p - 1 (p = 3, 4 \cdots)$. The infeed for each pass can then be computed and tabulated. The Programming and Meanings See Table 3-17.

程序员使用五次切削来加工螺纹。每次的径向进给从公式中获得：

$$\text{Infeed}_p = \frac{d}{\sqrt{p-1}} \times \sqrt{j}$$

其中，d 是牙深，p 是螺纹切削次数，第一次切削：$j=0.3$；第二次切削：$j=1$；$j=p-1(p=3,4\cdots)$ 然后可以计算并列表显示每次切削的进给量。指令及其含义见表3-17。

Table 3-17　The Programmings and Meanings

Programming pattern	Word address command	Meaning
	O2012	Program number
	(…)	Optional setup statement
	N0010 G21	Metric mode
	N0020 G50 X203 Z76	Set machining origin at①
	(TOOL1：THREAD AS PER PRINT)	Optional comment statement
Tool 1 in turret Machine start Up sequence	N0020 T0100	Return to tool change position①. Index to tool 1. Cancel values in offset register 1
	N0030 G0 M41	Shift to intermediate gear range
	N0040 G97 S200 M3	Cancel constant surface speed control. Spindle on (cw) at 200 r/min
	N0050 G0 X53.6 Z5 T0101 M8	Rapid tool to start point⑧. use offset values in register1. coolant on
Operation 1：Cut single Pass threats	N0060 X50.47	Rapid tool to first X depth position in metric
	N0070 G32 Z−20.5 F1.588	$X = 51 - 2 \times \text{Infeed}_1$
	N0080 G0 X53.6	$= 51 - 2 \times 0.974/\sqrt{4} \times \sqrt{0.3} = 50.47$
	N0090 Z5	Execute threading **pass1** and return to start point
	N0100 X50.03	Rapid tool to second X depth position in metric
	N0110 G32 Z−20.5 F1.588	$X = 51 - 2 \times \text{Infeed}_2$
	N0120 G0 X53.6	$= 51 - 2 \times 0.974/\sqrt{4} \times \sqrt{1} = 50.03$
	N0130 Z5	Execute threading **pass2** and return to start point
	N0140 X49.62	Rapid tool to third **X** depth position in metric
	N0150 G32 Z−20.5 F1.588	$X = 51 - 2 \times \text{Infeed}_3$
	N0160 G0 X53.6	$= 51 - 2 \times 0.974/\sqrt{4} \times \sqrt{2} = 49.62$
	N0170 G0 Z5	Execute threading **pass3** and return to start point
	N0180 X 49.31	Rapid tool to fourth **X** depth position in metric
	N0190 G32 Z−20.5 F1.588	$X = 51 - 2 \times \text{Infeed}_4$
	N0200 G0 X53.6	$= 51 - 2 \times 0.974/\sqrt{4} \times \sqrt{3} = 49.31$
	N0210 Z5	Execute threading **pass4** and return to start point
	N0220 X 49.05	Rapid tool to final **X** depth position in metric
	N0230 G32 Z−20.5 F1.588	$X = 51 - 2 \times \text{Infeed}_5$
	N0240 G0 X53.6	$= 51 - 2 \times 0.974/\sqrt{4} \times \sqrt{4} = 49.05$
	N0250 Z5	Execute threading **pass5** and return to start point

续表

Programming pattern	Word address command	Meaning
Machine stop Program end sequence	N0260 G0 X203 Z76 T0100 M9	Rapid to tool change position①. cancel values in offset register1. coolant off
	N0270 M5	Spindle off
	N0280 M30	Program end. Memory reset

表 3-17 指令及其含义

程序模块	指令	含义
	O2012	程序编号
	(…)	可选的设置数据
	N0010 G21	米制模式
	N0020 G50 X203 Z76	将机床原点设置为①
	(TOOL1: THREAD AS PER PRINT)	可选注释语句
机床启动过程,刀具1位于加工位	N0020 T0100	返回换刀位置,换1号刀,取消补偿文件1中的值
	N0030 G0 M41	主轴切换到低速挡
	N0040 G97 S200 M3	取消恒线速度,主轴以 200 r/min 正转
	N0050 G0 X53.6 Z5 T0101 M8	快速移动到⑧,使用补偿文件中1的值,打开冷却液
操作1: 切削螺纹	N0060 X 50.47	快速移刀到第一个 X 深度位置 $X = 51 - 2 \times \text{Infeed}_1$ $= 51 - 2 \times 0.974 / \sqrt{4} \times \sqrt{0.3} = 50.47$。
	N0070 G32 Z-20.5 F1.588	
	N0080 G0 X53.6	
	N0090 Z5	执行螺纹切削第一线程并返回到起始点
	N0100 X 50.03	快速移刀到第二个 X 深度位置 $X = 51 - 2 \times \text{Infeed}_2$ $= 51 - 2 \times 0.974 / \sqrt{4} \times \sqrt{1} = 50.03$
	N0110 G32 Z-20.5 F1.588	
	N0120 G0 X53.6	
	N0130 Z5	执行螺纹切削第二线程并返回到起始点
	N0140 X 49.62	快速移刀到第三个 X 深度位置 $X = 51 - 2 \times \text{Infeed}_3$ $= 51 - 2 \times 0.974 / \sqrt{4} \times \sqrt{2} = 49.62$,
	N0150 G32 Z-20.5 F1.588	
	N0160 G0 X53.6	
	N0170 G0 Z5	执行螺纹切削第三线程并返回到起始点
	N0180 X49.31	快速移刀到第四个 X 深度位置 $X = 51 - 2 \times \text{Infeed}_4$ $= 51 - 2 \times 0.974 / \sqrt{4} \times \sqrt{3} = 49.31$
	N0190 G32 Z-20.5 F1.588	
	N0200 G0 X53.6	
	N0210 Z5	执行螺纹切削第四线程并返回到起始点
	N0220 X49.05	快速移刀到第五个 X 深度位置 $X = 51 - 2 \times \text{Infeed}_5$ $= 51 - 2 \times 0.974 / \sqrt{4} \times \sqrt{4} = 49.05$
	N0230 G32 Z-20.5 F1.588	
	N0240 G0 X53.6	
	N0250 Z5	执行螺纹切削第五线程并返回到起始点

续表

程序模块	指令	含义
加工停止。程序结束	N0260 G0 X203 Z76T0100 M9	快速到换刀位置①.,取消值补偿文件 1 中的值,冷却液关闭
	N0270 M5	主轴关闭
	N0280 M30	程序结束。内存重置

3.4.2　Multiple-Pass Threading Cycle：G92

3.4.2　多线螺纹循环指令：G92

The G92 cycle reduces the number of programming blocks required for threading operations. With G92 threading, only one block of data is needed to execute one threading pass. As with G32, the tool must first be positioned at the start point prior to programming the cycle. See Figure 3-30.

G92 循环减少了螺纹切削操作所需的编程段数量。使用 G92 编程,只需要一个程序段来执行一次螺纹切削。与 G32 一样,在编程循环之前,必须首先将刀具定位在螺纹循环起点,如图 3-30 所示。

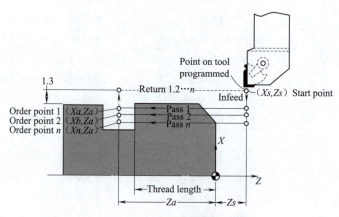

Figure 3-30　The Cutting Path of G92

图 3-30　G92 的切削路径

General Syntax

G92 Xn Zn Fn

G92 initiates a multiple-pass threading cycle.

Xn Zn Specifies the absolute X and Z absolute coordinates of the tool point after each threading pass (order point).

Fn Specifies the feed rate.

常用格式

G92 Xn Zn Fn

G92 启动多线螺纹车削循环。

Xn Zn 指每个线程切削后刀具的 X 和 Z 绝对坐标（螺纹切削终点）。

Fn 指进给量。

NOTES:

(1) For incremental coordinates replace X with U and Z with W. W and U indicate the direction and distance from the start point to the order point along the X and Z axes, respectively.

(2) The feed rate (thread lead) can have up to six digits accuracy to the right of the decimal point.

(3) Only straight and tapered can be cut. Outside as well as inside threading is permitted.

(4) The starting and stopping distances Zs, Za can be determined from the manufacturer's programming manual or estimated by the formulas:

$$Zs \approx (3 \text{ to } 4) \times \text{thread pitch}$$
$$Za \approx 0.5 \times Zs + \text{thread length}$$

(5) The infeed direction is straight in.

注意事项:

(1)对于增量坐标,用 U 代替 X,用 W 代替 Z。W 和 U 分别表示沿 X 轴和 Z 轴从螺纹循环起点到螺纹切削终点的方向和距离。

(2)进给量(螺纹导程)在小数点后最多可精确到六位数。

(3)只能切圆柱螺纹和锥形。允许外螺纹和内螺纹切削。

(4)起始距离 Zs 和终止距离 Za 可以从制造商的编程手册中确定或通过公式估算:

$$Zs \approx (3 \sim 4) \times 螺距$$
$$Za \approx 0.5 \times Zs + 螺纹长度$$

(5)进给方向是直线进给。

视频

G92螺纹切削单一循环指令2D轨迹

■ **EXAMPLE 3-13**

Using the G92 cycle, rewrite the program segment in Table 3-18 for the threading, as outlined in Example 3-12.

视频

G92螺纹车削循环指令3D视频

■ **例 3-13**

在表 3-18 中使用 G92 循环编写例 3-12 的程序段。

Table 3-18   The Programmings and Meanings

Programming pattern	Word address command	Meaning
	O2013	Program number
	(..........)	Optional setup statements
Tool 1 in turret Machine start Up sequence	N0010 G20	Metric mode
	N0020 G50 X203 Z76	Set machining origin at ①
	(TOOL1：THREAD AS PER PRINT)	Optional comment statement
	N0030 T0100	Return to tool change position①. Index to tool 1. Cancel values in offset register 1
	N0040 G0 M41	Shift to intermediate gear range
	N0050 G97 S200 M3	Cancel constant surface speed control. spindle on (cw) at 200 r/min
	N0060 G0X53.6 Z5 T0101 M8	Rapid tool to start point ⑧. use offset values in register1. coolant on
Operation 1： Cut multiple Pass threats	N0070 G92 X50.47 Z-20.5 F1.588	Begin multiple pass threading cycle; Rapid tool to first x depth. Thread to feed rate to order point① on the Z-axis. Return at rapid to start point. Pass1
	N0080 X50.03	Repeat cycle for pass2
	N0090 X49.62	Repeat cycle for pass3
	N0100 X49.31	Repeat cycle for pass4
	N0110 X49.05	Repeat cycle for pass5
	N0120 X49.05	Repeat G92 cycle for spring pass
Machine stop Program end sequence	N0130 G0 X203 Z76 T0100 M9	Rapid to tool change position①. cancel values in offset register1. coolant .off
	N0140 M5	Spindle off
	N0150 M30	Program end. Memory reset

G76螺纹切削
复合循环指令
2D轨迹

螺纹环规
使用方法
3D视频

螺纹塞规
使用方法
3D视频

表 3-18 指令及其含义

程序模块	指令	含义
	O2013	程序编号
	(..........)	可选的设置数据
机床启动过程 刀具 1 位于加工位置	N0010 G20	米制模式
	N0020 G50 X203 Z76	将机床原点设置为①
	（刀具 1：THREAD AS PER PRINT）	可选注释语句
	N0030 T0100	快速至换刀位置①，换 1 号刀，取消补偿文件 1 中的值
	N0040 G0 M41	主轴切换到低速挡
	N0050 G97 S200 M3	取消恒线速度。主轴以 200 r/min 正转
	N0060 G0 X53.6Z5T0101 M8	快速至循环起始点⑧，使用补偿文件 1 中的值，打开冷却液
操作 1：切削多线螺纹	N0070 G92 X50.47 Z−20.5 F1.588	开始 G96 循环，快速定位刀具到第一个 x 深度。螺纹进给量到 Z 轴上的顺序点①。快速返回循环起始点。线程 1
	N0080 X50.03	线程 2 的循环
	N0090 X49.62	线程 3 的循环
	N0100 X49.31	线程 4 的循环
	N0110 X49.05	线程 5 的循环
	N0120 X49.05	重复 G92 循环 Pass 5
加工停止 程序结束	N0130 G0 X203 Z76 T0100 M9	快速到换刀位置③，取消补偿 1 文件中的值，冷却液关闭
	N0140 M5	主轴关闭
	N0150 M30	程序结束。内存重置

3.4.3　Multiple Repetitive Threading Cycle：G76
3.4.3　多重复合螺纹循环指令：G76

The G76 cycle reduces the number of programming blocks involved in executing a threading operation. Several controllers require only a two-block set to be input. This cycle is capable of executing multiple threading passes at a specified infeed angle, but the programmer must first position the tool at an appropriate start point before entering the G76 word. See Figure 3-31.

G76 循环减少了执行螺纹车削操作所需的编程段数量。部分控制器仅需要输入两个程序段。该循环能够以指定的进给角度执行螺纹加工，但程序员必须首先将刀具定位在适当的循环起始点，然后才能使用 G76 指令，如图 3-31 所示。

视频
G76复合螺纹车削循环指令3D视频

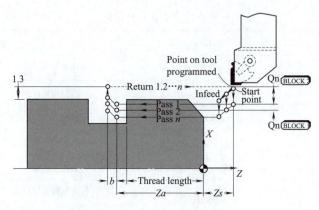

Figure 3-31　The Cutting Path of G76

图 3-31　G76 的加工路径

 **General Syntax**

G76Pabc Qn Rn(BLOCK1)

G76Xn Za Rn Pn Qn Fn (BLOCK2)

G76 initiates a multiple repitive threading cycle.

Pabc Contains three pairs of two-digit codes a, b, c

Numeric value of a can be 01, 02... or 99. Value specifics the number of finishing cuts.

Numeric value of b can be 00, 01... or 99. Value specifics the chamfer length of the pullout at the end of the cut.

Example: if b = 02

chamfer length = 2 × thread lead

Numeric value of c can be 00, 29, 55, 60, or 80. Value specifics the tool tip angle in degrees.

Qn(BLOCK1) Value of n specifics the minimum cutting depth per pass. Decimal point not accepted.

Rn (BLOCK1) Value of n specifics the finish allowance. Decimal point not accepted.

Xn The value of n specifics the thread minor diameter for OD thread or the major diameter for ID threads. See Figure 3-32.

Za The value of a specifies the end position of threads on the Z-axis.

Rn(BLOCK2) For tapered thread, the value of n specifies the difference in the thread radius between the start and end positions of the thread at the final pass. For straight cut thread the value of n is zero and R can be omitted.

Pn(BLOCK2) The value of n specifies the final thread depth per side. Decimal point not accepted. See Figure 3-32.

Qn (BLOCK2) The value of n specifies the depth per side of the first cut. Decimal point not accepted. See Figure 3-32.

Fn The value of n specifics the feed rate. Note the feed rate is equal to the thread pitch for common single start threads.

常用格式

G76Pabc Qn Rn(BLOCK1)

G76Xn Za Rn Pn Qn Fn (BLOCK2)

G76 执行螺纹加工复合循环指令。

Pabc 包含三对两位数代码 a,b,c。

a 可以是 01、02 ……或 99,具体表示精加工切削的数量。

b 可以是 00、01 ……或 99,具体表示切削结束时加工的倒角长度。

例:如果 b = 02

倒角长度 = 2 × 螺纹导程

c 可以是 00、29、55、60 或 80,以度为单位指定刀尖角度。

Qn(程序段 1) n 是每次的最小切削深度,没有小数点。

Rn(程序段 1)n 指精车余量,没有小数点。

Xn n 表示外螺纹的螺纹小径或内螺纹的大径,如图 3-32 所示。

Za a 指 Z 轴上螺纹的最终位置。

Rn(程序段 2)对于锥形螺纹,n 指在最后一次切削中,加工过程中螺纹循环起始点和切削终点位置之间的螺纹半径差,n 的值为零,可以省略 R。

Pn(程序段 2)n 指的螺纹牙深,没有小数点,如图 3-32 所示。

Qn(程序段 2)n 的值指第一次切削深度,没有小数点,如图 3-32 所示。

Fn n 表示进给量。请注意,进给量等于单线螺纹的螺距。

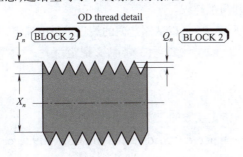

Figure 3-32 Processing example

图 3-32 加工实例

NOTES:

(1) For incremental coordinates replace X with U and Z with W. W and U indicate the direction and distance from the start point to the order point along the X and Z axes, respectively.

(2) The feed rate (thread lead) can have up to six digits accuracy to the right of the decimal point.

(3) Only straight and tapered can be cut.

(4) The starting and stopping distances Zs, Za can be determined from the manufacturer's programming manual or estimated by the formulas:

$$Zs \approx (3 \text{ to } 4) \times \text{thread pitch}$$
$$Za \approx 0.5 \times Zs + \text{thread length}$$

(5) The infeed direction is straight in. See Figure 3-33.

(6) The angle at which infeed occurs will be half the tool tip angle specified by the value c. The default tool tip value is 60.

视频

梯形螺纹加工方法

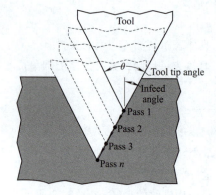

Figure 3-33　Processing example

图 3-33　加工实例

注意事项：

（1）对于增量坐标，用 U 代替 X，用 W 代替 Z。W 和 U 分别表示沿 X 轴和 Z 轴从循环起点到切削终点的方向和距离。

（2）进给量（螺纹导程）在小数点后最多可精确到六位数。

（3）只能切削圆柱螺纹和圆锥螺纹。

（4）起始距离 Zs 和终止距离 Za 可以从制造商的编程手册中确定，也可以根据公式估算：

$$Zs \approx (3 \text{ 到 } 4) \times \text{螺距}$$
$$Za \approx 0.5 \times Zs + \text{螺纹长度}$$

（5）进给方向是直线进刀，如图 3-33 所示。

（6）进刀的角度将是值 c 指定度数的一半。默认值为 60。

It is possible to machine tapered threads using the G32, G92, or G76 cycles. The CNC machine programming manual will supply the information for such operations.

可以使用 G32、G92 或 G76 循环加工圆锥螺纹，数控机床编程手册将提供此类操作的信息。

■ **EXAMPLE 3-14**

Execute the threading operation of Example 3-12, by using a G76 cycle writtern in Table 3-19.

■ **例 3-14**

使用 G76 循环完成例 3-12，将程序写在表 3-19 中。

Chapter 3 Techniques and Fixed Cycles for CNC Lathe Programming

Table 3-19 The Programmings and Meanings

Programming pattern	Word address command	Meaning
	O2014	Program number
	(..........)	Optional setup statements
Tool 1 in turret Machine start Up sequence	N0010 G20	Metric mode
	N0020 G50 X203Z76	Set machining origin at ①
	(TOOL1: THREAD AS PER PRINT)	Optional comment statement
	N0030 T0100	Return to tool change position①. Index to tool 1. Cancel values in offset register 1
	N0040 G0 M41	Shift to intermediate gear range
	N0050 G97 S200 M3	Cancel constant surface speed control. spindle on (cw) at 200 r/min
	N0060 G0 X53.6 Z5T0101 M8	Rapid tool to start point ⑧. use offset values in register1. coolant on
Operation 1: Cut multiple Pass threats	N0070 G76 P010060 Q050 R200	Begin executing BLOCK1 of the multiple-pass threating cycle
	N0080 G76 X49.05 Z−20.5P840 Q240 F1.588	Execute BLOCK2 of the threading cycle
Machine stop Program end sequence	N0090 G0X203 Z76 T0100 M9	Rapid to tool change position③. cancel values in offset register1. coolant off
	N0100 M5	Spindle off
	N0110 M30	Program end. Memory reset

表 3-19 指令及其含义

编程模式	指令	含义
	O2014	程序编号
	(..........)	可选的设置数据
机床启动过程	N0010 G20	米制模式
	N0020 G50 X203Z76	将机床原点设置为①
	(TOOL1: THREAD AS PER PRINT)	可选注释语句
	N0030 T0100	返回换刀位置①,换1号刀,取消补偿文件1中的值
	N0040 G0 M41	主轴切换到低速挡
	N0050 G97 S200 M3	取消恒线速度,主轴以200 r/min 正转
	N0060 G0 X53.6 Z5T0101 M8	快速运动到循环起始点⑧,使用补偿文件1中的值,打开冷却液
操作1:切削螺纹	N0070 G76 P010060 Q050 R200	执行螺纹切削复合循环指令的第一行程序
	N0080 G76 X49.05 Z−20.5P840 Q240 F1.5	执行螺纹切削复合循环指令的第二行程序
加工停止, 程序结束	N0090 G0 X203 Z76 T0100 M9	快速换到换刀位置③,取消在补偿文件1中的值。冷却液关闭
	N0100 M5	主轴关闭
	N0110 M30	程序结束。内存重置

Chapter Summary

本章总结

The following key concepts were discussed in this chapter:

(1) *Tool edge programming involves programming the part geometry directly. This technique can only be used to machine horizontal or vertical lines.*

(2) *Tool nose radius (TNR) compensation involves programming the part geometry directly instead of the center of the tool nose radius. The part geometry can include arcs and tapered lines.*

(3) *TNR compensation is initiated or canceled by the next linear (G0 or G1) tool motion command following a G41, G42, or G40 word.*

(4) *The setup person prepares the CNC lathe for TNR programming by keying in the following information*:

- *X and Z tool offsets to the tool edge.*
- *Tool nose radius value.*
- *Tool nose vector number.*

(5) *The tool nose vector indicates to the controller the direction of the tool edge from the TNR center.*

(6) *The G90 cycle is used to rough turn or bore a part from stock. Straight or tapered passes may be programmed.*

(7) *Rough facing from stock can be executed via the G94 cycle. Straight or tapered passes may be programmed.*

(8) *The multiple repetitive cycle G71 is used to rough cut general profiles consisting of arcs and tapered lines. This can be followed by a G70 finishing pass cycle.*

(9) *The following cycles can be used for threading*:

G32——single-pass cycle (four blocks programmed per pass).

G92——multiple-pass cycle (one block programmed per pass).

G76——multiple-pass cycle (two blocks programmed for all passes).

本章讨论了以下关键概念：

(1) 刀具刀尖编程可直接编写零件轮廓程序。此程序只能用于加工水平或垂直线。

(2) 刀尖圆弧半径(TNR)补偿可直接使用加工零件轮廓进行编程，而不考虑刀尖圆弧的圆心。零件轮廓包括弧形和锥形线。

(3) 在G41、G42或G40后的下一个线性(G0或G1)移动中启动或取消刀尖圆弧半径补偿。

(4) 操作工通过输入以下信息为数控车床编程准备刀尖圆弧半径：

- 刀具刀尖 X 轴和 Z 轴补偿值。
- 刀尖圆弧半径值。
- 刀尖矢量值。

(5) 刀尖矢量向控制器指示刀具刀尖位于刀尖圆弧圆心的方向。

(6) G90 循环用于粗加工外圆或内孔,可以编程直线或斜线。

(7) 端面粗加工可以通过 G94 循环执行,可以编程直线或斜线。

(8) 复合循环指令 G71 用于粗加工由弧和锥形线组成的一般轮廓,然后使用 G70 指令进行精加工循环。

(9) 以下循环可用于加工螺纹:

G32——单线循环指令(每线程使用四条程序段)。

G92——多线螺纹循环指令(每线程使用一条程序段)。

G76——多重复合螺纹循环指令(一条螺纹使用两个程序段)。

Review Exercises

回顾练习

3.1 Write a word address program segment for rough turning the part shown in Figure 3-34. The machining is documented in the CNC Tool and Operations Sheet below (Table 3-20). The profile consists only of horizontal and vertical lines, so tool edge programming can be applied.

Program number: O2015.

Notes:

Limit the rotation of the spindle to 3 000 r/min when G96 is in effect.

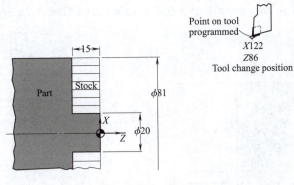

Figure 3-34 Case 3.1

图 3-34 例 3.1

Table 3-20　CNC Tool and Operations Sheet

Material: Al Alloy 7075 - T6							
Tool	Offset	Operation	Speed (m/min)	Feed (mm/r)	Tool Description	Insert	TNR
1	01	Rough turn OD Use a G90 roughing cycle. Remove 1.5 mm per side for each pass.	500	0.3	Turning holder MCLNR 12 - 4	CNMG - 432 - NR	0.8

3.1 写一段程序，对图 3-34 所示的零件进行粗加工。加工信息在数控刀具和加工（表 3-20）中。轮廓仅由水平和垂直线组成，因此可以应用刀具刀尖编程。

程序编号：O2015。

注意事项：

当 G96 生效时，将主轴的转速限制为 3 000 r/min。

表 3-20　数控刀具和加工卡

材质：铝合金 7075 - T6							
刀具号	补偿号	操作	主轴转速 (m/min)	进给量 (mm/r)	刀具描述	刀片	TNR
1	01	粗车外圆，使用 G90 粗加工循环，粗车每次切削深度为 1.5 mm	500	0.3	外圆车刀刀杆 MCLNR 12 - 4	CNMG - 432 - NR	0.8

3.2 The part shown in Figure 3-35 is to be rough faced from stock. Use the CNC Tool and Operations Sheet (Table 3-21) given below to write the word address program segment. Tool edge programming may be applied since the profile consists only of horizontal and vertical lines.

Program number 02016.

Notes：

Limit the rotation of the spindle to 3 000 r/min when G96 is in effect.

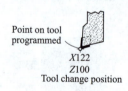

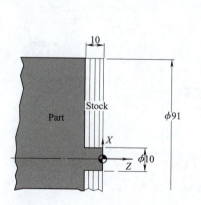

Figure 3-35　Case 3.2

图 3-35　例 3.2

Table 3-21   CNC Tool and Operations Sheet

Material: Al Alloy 7075 – T6

Tool	Offset	Operation	Speed (m/min)	Feed (mm/r)	Tool Description	Insert	TNR
1	01	Rough turn OD, Use a G94 roughing cycle. Remove 0.5 mm per side for each facing pass	500	0.2	Turning holder MCLNR 12 – 4	CNMG – 432 – NR	0.8

3.2 图 3-35 中所示的零件需要进行端面粗加工。使用下面给出的数控刀具和加工表编写程序段(表 3-21)。可以应用刀具刀尖编程,因为轮廓仅由水平和垂直线组成。

程序号 O2016。

注意事项:

当 G96 生效时,将主轴的转速限制为 3 000 r/min。

表 3-21   数控刀具和加工表

材质: 铝合金 7075 – T6

刀具号	补偿号	操作	主轴转速 (m/min)	进给量 (mm/r)	刀具描述	刀片	TNR
1	01	粗车外圆,使用 G94 粗加工循环,每次切削深度 为 0.5 mm	500	0.2	外圆车刀刀杆 MCLNR 12 – 4	CNMG – 432 – NR	0.8

3.3 The profile shown in Figure 3-36 is to be machined from stock. Write a word address program to execute the rough and finish turning as described in the CNC Tool and Operations Sheet(Table 3-22). The setup person has entered the tool offsets for tool edge programming, tool nose radius values, and tool nose vector 3 into the offset registers.

Program number: O2017.

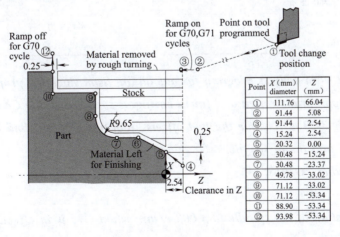

Figure 3-36   Case 3.3

图 3-36   例 3.3

Notes:

Limit the rotation of the spindle to 3000 r/min when G96 is in effect.

Table 3-22 CNC Tool and Operations Sheet

Material: Al Alloy 7075 – T6

Tool	Offset	Operation	Speed (m/min)	Feed (mm/r)	Tool Description	Insert	TNR
1	01	Rough turn OD, Use a G71 roughing cycle. Remove 0.5 mm per side for each pass. Leave 0.25 mm for finishing	600	0.2	外圆车刀刀杆 MCLNR 12 – 4	CNMG – 432 – NR	0.8
2	02	Finish ID contour, Use a G70 finishing cycle	800	0.1	外圆车刀刀杆 MCLNR 12 – 4	CNMG – 434 – NF	1.6

3.3 图 3-36 所示的零件需要进行外圆粗加工。根据下表（表 3-22）编写程序来执行粗加工和精加工车削。操作工已将刀具刀尖编程的刀具补偿、刀尖圆弧半径值和刀尖矢量 3 输入到补偿文件中。

程序编号：O2017。

注意事项：

G96 生效时，将主轴的转速限制为 3 000 r/min。

表 3-22 数控刀具和加工表

材质：铝合金 7075 – T6

刀具号	补偿号	操作	主轴转速 (m/min)	进给量 (mm/r)	刀具描述	刀片	TNR
1	01	粗车外圆，G71 粗加工循环，每次切削深度为 0.5 mm，精车余量为 0.25 mm	600	0.2	外圆车刀刀杆 MCLNR 12 – 4	CNMG – 432 – NR	0.8
2	02	精车外圆，用 G70 精加工循环	800	0.1	外圆车刀刀杆 MCLNR 12 – 4	CNMG – 434 – NF	1.6

3.4 Write a word address program for machining from stock the profile shown in Figure 3-37. Apply appropriate rough and finish boring as specified in the CNC Tool and Operations Sheet (Table 3-23). Assume the tool offsets, tool nose radius, and tool nose vector 2 have been entered into the offset file.

Program number: O2018.

Notes:

Limit the rotation of the spindle to 3 000 r/min when G96 is in effect.

Chapter 3 Techniques and Fixed Cycles for CNC Lathe Programming

第 3 章 车削固定循环指令

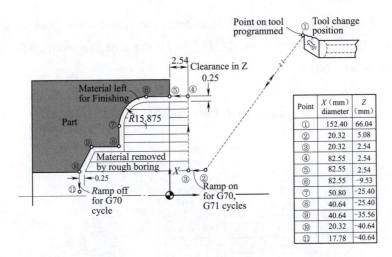

Figure 3-37 Case 3.4

图 3-37 例 3.4

Table 3-23 CNC Tool and Operations Sheet

Tool	Offset	Operation	Speed (m/min)	Feed (mm/r)	Tool Description	Insert	TNR
Material: Al Alloy 7075-T6							
1	01	Rough turn ID, Use a G71 roughing cycle. Remove 0.5 mm per side for each pass. Leave 0.25 mm for finishing	500	0.3	Boring holder S-MCLNR 20-4	CNMG-432-NR	0.8
2	02	Finish ID contour, Use a G70 finishing cycle	600	0.07	Boring holder S-MCLNR 20-4	CNMG-434-NF	1.6

3.4 编写一段程序，用于加工如图 3-37 所示的形状。根据下表 3-23 编写程序来执行粗加工和精加工车削。假设刀具补偿、刀尖半径和刀尖矢量 2 已输入到补偿文件中。

程序编号：O2018。

注意事项：

当 G96 生效时，将主轴的转速限制为 3 000 r/min。

表 3-23 数控刀具和加工表

刀具号	补偿号	操作	主轴转速 (m/min)	进给量 (mm/r)	刀具描述	刀片	TNR
材质：铝合金 7075-T6							
1	01	粗车内孔，G71 粗加工循环，每次切削深度为 0.5 mm，精车余量为 0.25 mm	500	0.3	镗刀杆 S-MCLNR 20-4	CNMG-432-NR	0.8
2	02	精车内孔，用 G70 精加工循环	600	0.07	镗刀杆 S-MCLNR 20-4	CNMG-434-NF	1.6

3.5 The profile shown in Figure 3-38 is to be machined from stock as outlined in the CNC Tool and Operations Sheet(Table 3-24). Write the required word address program.

Program number: O2019.

Notes:

Limit the rotation of the spindle to 3 000 r/min when G96 is in effect.

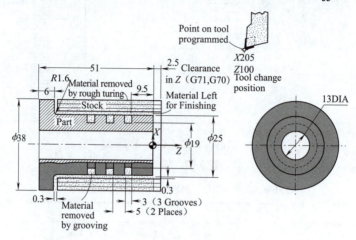

Figure 3-38   Case 3.5

图 3-38   例 3.5

Table 3-24   CNC Tool and Operations Sheet

Material: Al Alloy 7075 - T6

Tool	Offset	Operation	Speed (m/min)	Feed (mm/r)	Tool Description	Insert	TNR
1	01	Rough turn OD, Use a G71 roughing cycle. Remove 0.5 mm per side for each pass. Leave 0.3 mm for finishing	600	0.2	Turning holder MCLNR 12 - 4	CNMG - 432 - NR	0.8
2	02	Finish OD contour, Use a G70 finishing cycle	800	0.05	Turning holder MCLNR 12 - 4	CNMG - 434 - NF	1.6
3	03	Peck drill right end Use a G74 peck drilling cycle. Set the peck depth at 4mm	800	0.3	13 dia. drill		
4	04	Groove OD, Use a G75 grooving cycle. Set the peck depth at 1mm	400	0.05	DGTR 19 - 3	DGNM 3202J	0.2

3.5 图 3-38 中所示的轮廓应按照表 3-24 数控刀具和操作表中的说明进行加工。写下所需的字地址程序。

程序编号:O2019。

注意事项:

当 G96 生效时,将主轴的转速限制为 3 000 r/min。

表3-24 数控刀具和加工表

材质:铝合金 7075-T6

刀具号	补偿号	操作	主轴转速(m/min)	进给量(mm/r)	刀具描述	刀片	TNR
1	01	粗车外圆,用 G71 粗加工循环,每次切削深度为 0.5 mm,留下 0.3 mm 用于精加工	600	0.2	外圆车刀刀杆 MCLNR 12-4	CNMG-432-NR	0.8
2	02	精车外圆,用 G70 精加工循环	800	0.05	外圆车刀刀杆 MCLNR 12-4	CNMG-434-NF	1.6
3	03	钻右端孔,使用 G74 循环,将每次切削深度设置为 4 mm	800	0.3	φ13 的钻头 Point on tool programmed		
4	04	切槽,使用 G75 循环。将每次切削深度设置为 1 mm	400	0.05	切断刀杆 19-3 Point on tool programmed	DGNM 3202J	0.2

3.6 The part shown in Figure 3-39 is to be rough and finish turned from stock. Face grooves are cut into the right end to complete the machining. Use the information provided in the CNC Tool and Operations Sheet(Table 3-25)to write the required word address program.

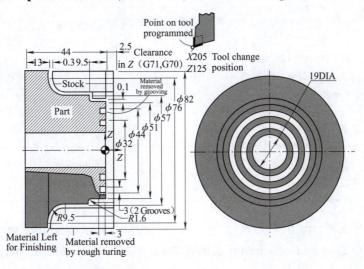

Figure 3-39 Case 3.6

图 3-39 例 3.6

Program number: O2020.

Notes:

Limit the rotation of the spindle to 3 000 r/min when G96 is in effect.

Table 3-25 CNC Tool and Operations Sheet

Material: Al Alloy 7075 – T6

Tool	Offset	Operation	Speed (m/min)	Feed (mm/r)	Tool Description	Insert	TNR
1	01	Rough turn OD, Use a G71 roughing cycle. Remove 0.5 mm per side for each pass. Leave 0.1 mm for finishing	650	0.3	Turning holder MCLNR 12 – 4	CNMG – 432 – NR	0.8
2	02	Finish OD contour, Use a G70 finishing cycle	800	0.05	Turning holder MCLNR 12 – 4	CNMG – 434 – NF	1.6
3	03	Face groove right end Use a G74 grooving cycle. Set the peck depth at 1 mm	300	0.05	TLSR – 1238	TLR 3125R	0.1

3.6 完成图 3-39 中所示的零件外圆粗加工和精加工和右端面槽的加工。使用表 3-25 数控刀具和操作表中提供的信息编写所需的程序。

程序编号：O2020。

注意事项：

当 G96 生效时，将主轴的转速限制为 3 000 r/min。

表 3-25 数控刀具和加工表

材质：铝合金 7075 – T6

刀具号	补偿号	操作	主轴转速 (m/min)	进给量 (mm/r)	刀具描述	刀片	TNR
1	01	粗车外圆，用 G71 粗加工循环，每次切削深度 0.5 mm，留下 0.1 mm 用于精加工	650	0.3	外圆车刀刀杆 MCLNR 12 – 4	CNMG – 432 – NR	0.8
2	02	精车外圆，用 G70 精加工循环	800	0.05	外圆车刀刀杆 MCLNR 12 – 4	CNMG – 434 – NF	1.6
3	03	切槽，使用 G74 切槽循环。每次切削深度 1 mm	300	0.05	端面槽刀杆 TLSR – 1238	TLR 3125R	0.1

3.7 The part shown in Figure 3-40 is to be produced by executing the operations outlined in the CNC Tool and Operations Sheet (Table 3-26) given below. Write the word address part program to machine the part.

Program number: O2021.

Chapter 3 Techniques and Fixed Cycles for CNC Lathe Programming

Notes:

Limit the rotation of the spindle to 3 000 r/min when G96 is in effect.

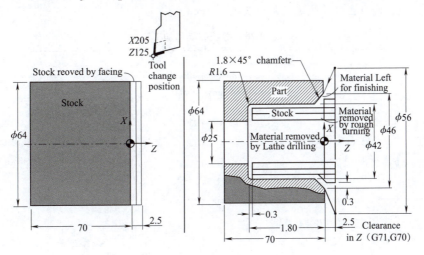

Figure 3-40　Case 46

图 3-40　例 46

Table 3-26　CNC Tool and Operations Sheet

Material: Al Alloy 7075 – T6

Tool	Offset	Operation	Speed (m/min)	Feed (mm/r)	Tool Description	Insert	TNR
1	01	Face end, Use a G94 facing cycle. Remove 1.3 mm per side for each facing pass	550	0.1	Turning holder MCLNR 12 – 4 Point on tool programmed	CNMG – 432 – NR	0.8
2	02	Peck drill right end Use a G74 peck drilling cycle. Set the peck depth at 5 mm	400	0.2	38 dia drill Point on tool programmed		
3	03	Rough bore ID, Use a G71 roughing cycle. Remove 0.5 mm per side for each pass. Leave 0.3 mm for finishing	400	0.1	Boring holder MCLNR 20 – 4 Point on tool programmed	CNMG – 432 – NR	0.8
4	04	Finish ID contour Use a G70 finishing cycle	450	0.05	Boring holder S-MCLNR 20 – 4	CNMG – 434 – NF	1.6

3.7　图 3-40 中所示的部件将通过执行下面给出的数据刀具和操作表 3-26 中列出的操作来完成。编写零件程序来加工零件。

程序编号：O2021。

注意事项：

当 G96 生效时，将主轴的转速限制为 3 000 r/min。

表 3-26 数控刀具和加工表

材质：铝合金 7075-T6

刀具号	补偿号	操作	主轴转速（m/min）	进给量（mm/r）	刀具描述	刀片	TNR
1	01	车端面，用 G94 循环，每次切削深度 1.3 mm	550	0.1	外圆车刀刀杆 MCLNR 12-4 Point on tool programmed	CNMG-432-NR	0.8
2	02	钻右孔，使用 G74 钻孔循环，每次切削深度为 5 mm	400	0.2	φ38 钻头 Point on tool programmed		
3	03	车内孔，使用 G71 粗加工循环。每次切削深度 0.5 mm。留下 0.3 mm 用于精加工	400	0.1	镗刀杆 MCLNR 20-4 Point on tool programmed	CNMG-432-NR	0.8
4	04	精车外圆，用 G70 精加工循环	450	0.05	镗刀杆 S-MCLNR 20-4	CNMG-434-NF	1.6

3.8 Write a word address program segment for threading the blank as shown in Figure 3-41. The machining operations are specified in the CNC Tool and Operations Sheet (Table 3-27) below.

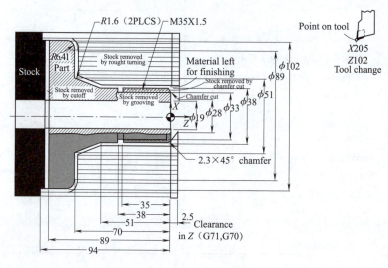

Figure 3-41 Case 3.8

图 3-41 例 3.8

Program number: O2023.

Notes:

Limit the rotation of the spindle to 3 000 r/min when G96 is in effect.

Table 3-27　CNC Tool and Operations Sheet

Material: Al Alloy 7075 – T6

Tool	Offset	Operation	Speed (m/min)	Feed (mm/r)	Tool Description	Insert	TNR
1	01	Rough turn OD, Use a G71 roughing cycle. Remove 2.5 mm per side for each pass. Leave 0.2 mm for finishing	120	0.2	TURNING HOLDER MCLNR 12 – 4	CNMG – 432 – NR	0.8
2	02	Finish OD contour, Use a G70 finishing cycle.	250	0.05	TURNING HOLDER MCLNR 12 – 4	CNMG – 434 – NF	1.6
3	03	Groove OD, Use a G75 grooving cycle. Set the peck depth at 1 mm	120	0.05	DGTR 19 – 3	DGNM 3202J	0.2
4	04	Cut threads, Use a G32 single-pass cycle. Set the number passes as: $p=5$ Set the starting distance as: $Zs = 4 \times$ thread pitch Set the stopping distance as: $Za = 0.5 \times Zs$ + thread length	350	THREAD PITCH	SER 0750 K16	22ER 6 UN	0.5
4	02	Chamfer end, Use a G1 Linear cut	300	0.05	TURNING HOLDER MCLNR 12 – 4	CNMG – 434 – NF	1.6
5	05	Peck cutoff, Use a G75 peck cutoff cycle	100	0.1	DO-GRIP BLADE DGFH 26 – 3	DGNM 320C	0.2

3.8　写一个程序段完成如图3-41所示的轮廓。加工操作如表3-27数控刀具和操作表所示。

程序编号：O2023。

注意事项：

当G96生效时，将主轴的转速限制在3 000 r/min。

表3-27　数控刀具和加工表

材质：铝合金7075 – T6

刀具号	补偿号	操作	主轴速度 (m/min)	进给量 (mm/r)	刀具描述	刀片	TNR
1	01	粗车外圆 用G71粗加工循环，每次切削深度2.5 mm，留下0.2 mm用于精加工	120	0.2	外圆车刀刀杆 MCLNR 12 – 4	CNMG – 432 – NR	0.8

续表

材质：铝合金 7075-T6

刀具号	补偿号	操作	主轴速度 (m/min)	进给量 (mm/r)	刀具描述	刀片	TNR
2	02	精车外轮廓，用 G70 精加工循环	250	0.0	外圆车刀刀杆 MCLNR 12-4	CNMG-434-NF	1.6
3	03	切槽，使用 G75 循环，将切削深度设置为 1 mm	120	0.05	切断刀杆 19-3 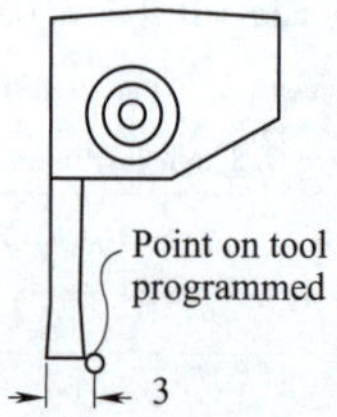	DGNM 3202J	0.2
4	04	车螺纹，使用 G32 单线循环。螺纹切削线程次数：$p=5$ 将起始距离设置为：$Zs=4\times$螺距 将终止距离设置为：$Za=10.5\times Zs+$螺纹长度	350	THREAD PITCH	外螺纹车刀 0750 K16 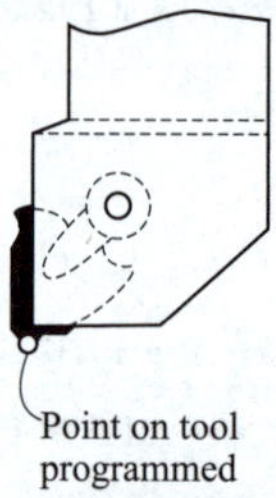	22ER 6 UN	0.5
4	02	车倒角，使用 G1 线性切削	300	0.05	外圆车刀刀杆 MCLNR 12-4	CNMG-434-NF	1.6
5	05	切断，用 G75 循环	100	1	切断刀杆 26-3	DGNM 320C	0.2

3.9 Use the G76 multiple repetitive cycle to cut the threads as specified in the CNC Tool and Operations Sheet in Exercise 3.8.

Program number：O2024.

When using the Predator simulator, input X in G76 as the diameter at which thread cutting starts.

3.9 按照习题3.8中的数控刀具和加工表中的规定，使用G76多次重复循环指令切削螺纹。

程序编号：O2024。

使用Predator模拟器时，在G76中输入X作为螺纹切削开始的直径。

Chapter 4 CNC Machining and Programming

第 4 章 数控加工及程序编制

Learning Objectives

At the conclusion of this chapter you will be able to:

(1) Understand the programming features and coordinate system of CNC lathes.

(2) Master the F/S/T function instructions of the FANUC system.

(3) Master G codes-G0/G1/G2/G3/G90/G94/G71/G73/G70.

(4) Master the common auxiliary function word M instruction.

(5) Master the manual programming method of shaft parts.

(6) Be able to formulate the processing scheme of shaft parts according to the structural characteristics and technical requirements of the parts.

(7) Be able to manually program parts.

学习目标

在本章的结尾你将能够：

(1) 了解数控车床的编程特点及坐标系统。

(2) 掌握 FANUC 系统的 F/S/T 功能指令。

(3) 掌握 G0/G1/G2/G3/G90/G94/G71/G73/G70 指令。

(4) 掌握常用辅助功能字 M 指令。

(5) 掌握轴类零件的手工编程方法。

(6) 能根据零件结构特点和技术要求，制订轴类零件的加工工艺方案。

(7) 能手工编制零件数控加工程序。

视频
零件自动加工过程

This chapter introduces the specific parts processing, mainly for project-based teaching. Through the process analysis and programming of specific parts, combined with the actual processing the learned theory to deepen the learning impression of the learners.

本章对具体的零件加工进行介绍，主要进行项目化教学。通过对具体零件的工艺分析及编程，将所学到的理论知识与实际加工相结合，加深学习者的学习印象，达到学以致用的效果。

4.1　Project 1　Program External Contour

4.1　项目1　外轮廓的数控编程

4.1.1　Project Import

4.1.1　项目导入

As shown in Figure 4-1, the bar is 45 steel and $\phi 52$. Write a program for finishing.

图 4-1 所示为典型的轴类零件,材料为 45#钢,毛坯为 $\phi 52$ 棒料,编写该零件的精加工程序。

The F, S, T codes, G0, G1, G2/G3 and other words are used to complete the programming of the external contour finishing program for shaft parts.

使用 F、S 和 T 指令,基本的运动指令 G0、G1 和 G2/G3 等完成轴类零件外圆轮廓精加工程序的编制。

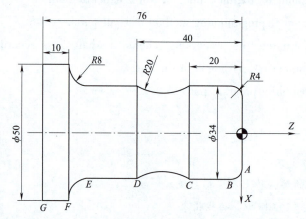

Figure 4-1　Case 4.1

图 4-1　例 4.1

4.1.2　Project Implementation

4.1.2　项目实施

1. Cutting Route

1. 走刀路线

Finish OD from right to left. The cutting route is $O \rightarrow J \rightarrow A \rightarrow B \rightarrow C \rightarrow D \rightarrow E \rightarrow F \rightarrow G \rightarrow H \rightarrow O$. See Figure 4-2.

精车时,从右至左精车整个外圆表面及倒角。以 $O \rightarrow J \rightarrow A \rightarrow B \rightarrow C \rightarrow D \rightarrow E \rightarrow F \rightarrow G \rightarrow H \rightarrow O$ 的路线精车外轮廓,走刀路线如图 4-2 所示。

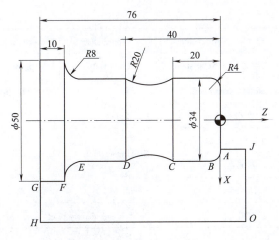

Figure 4-2 The Cutting Path
图 4-2 走刀路线

2. Program
2. 数控加工程序

The program origin is placed at the center of the right end face of the workpiece. The start point of circle is point O ($X54$, $Z2$). This point is usually closer to the workpiece, but there is a certain distance from the blank. The tool change point is set at ($X100$, $Z100$). The machining program is shown in Table 4-1.

工件坐标系原点放在工件右端面中心。车外圆起点为 O，坐标定为($X54,Z2$)，该点通常离工件较近，但与毛坯留有一定的距离。换刀点设在($X100,Z100$)处。数控加工程序见表 4-1。

Table 4-1 The Programming and Meaning

Part Number		Part Name	Stepped shaft	The program origin	Center of right face
Program number	O0201	CNC system	FANUC 0iT	prepared	
Block number	World address command			Meaning	
N10	T0101;			Index to tool 1, use offset values in offset file 1	
N20	G97 G99M3 S600;			Cancel constant surface speed control, spindle on (CW) at 600 r/min	
N30	G0 X54 Z2;			Rapid move to the start point O (54.2)	
N40	X26;			Rapid move to J	
N50	Z0;			Rapid move to A	
N60	G3 X34 Z-4 R4 F0.2;			Cut to B at feedrate	
N70	G1 W-16;			Cut to C	
N80	G2 X34 Z-40 R20;			Cut to D	
N90	G1 Z-58;			Cut to E	
N100	G2 X50 Z-66 R8;			Cut to F	
N110	G1 Z-76;			Cut to G	
N140	X54;			Cut to H	

续表

Part Number		Part Name	Stepped shaft	The program origin prepared	Center of right face
Program number	O0201	CNC system	FANUC 0iT		
Block number	World address command		Meaning		
N150	G0 Z2;		Rapid to O		
N160	G0 X100;		Rapid move to the tool change point		
N170	Z100;				
N180	M5;		Spindle off		
N190	M30;		Program end Memory reset		

表 4-1 数控加工程序单

零件号		零件名称	阶梯轴	编程原点	右端面中心
程序号	O0201	数控系统	FANUC 0iT	编制	
程序段号	程序内容		程序说明		
N10	T0101;		选择 1 号刀,刀补号为 1 号		
N20	G97 G99 M3 S600;		主轴正转,设定粗车转速为 600 rpm		
N30	G0 X54 Z2;		刀具快速定位到工件附近 O(54.2)		
N40	X26;		快速移动到 J 点		
N50	Z0;		快速移动到 A 点		
N60	G3 X34 Z-4 R4 F0.2;		使用进给量切削到 B 点		
N70	G1 W-16;		使用进给量切削到 C 点		
N80	G2 X34 Z-40 R20;		切到 D 点		
N90	G1 Z-58;		切到 E 点		
N100	G2 X50 Z-66 R8;		切到 F 点		
N110	G1 Z-76;		切到 G 点		
N140	X54;		切到 H 点		
N150	G0 Z2;		快速移动到 O 点		
N160	G0 X100;		快速退刀到换刀点		
N170	Z100;				
N180	M5;		主轴停止		
N190	M30;		程序结束,重置内存		

4.2 Project 2 Program a Simple Stepped Shaft

4.2 项目 2 阶梯轴的数控编程

4.2.1 Project Import

4.2.1 项目导入

As shown in Figure 4-3, the bar is 45 steel and $\phi 40$. Try to write the processing technology and the program for machine the part.

如图 4-3 所示阶梯轴,材料为 45#钢,毛坯为 $\phi 40$ 棒料,试制订零件的加工工艺,编写该零件的数控加工程序。

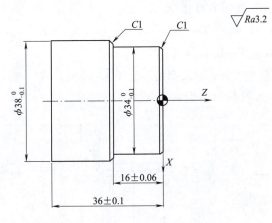

Figure 4-3　Case 4.2

图 4-3　例 4.2

4.2.2 Project Implementation

4.2.2 项目实施

1. Processing Technology

1. 零件加工工艺

(1) Part Drawing

(1) 零件图

As shown in Figure 4-3, this part has one step and two chamfers. The requiement of radial dimension accuracy and the surface accuracy are not high, and the roughness value is not larger than Ra 3.2.

如图 4-3 所示阶梯轴,该零件有一个台阶、两处倒角,径向尺寸精度要求不高,表面精度不高,表面粗糙度不大于 Ra 3.2。

（2）Clamping Method

（2）装夹方案

Choose a universal fixture：three-jaw chuck

选择通用夹具——三爪卡盘

（3）Processing Sequence and Cutting Path

（3）加工顺序及走刀路线

Step 1：Machine the end surface，the cutting path is $A \to H \to O \to I \to A$. See in Figure 4-4（a）. G94 facing is used.

工步1：车端面，采用$A \to H \to O \to I \to A$走刀路线，如图4-4(a)所示，使用G94指令。

Step 2：Rough turn OD. The route is $A \to B \to C \to D \to A \to E \to F \to G \to A$. Leave 0.1（in diameter）for finish. The cutting path is shown in Figure 4-4（b）. G90 cycle is used.

工步2：粗车外圆。粗加工采用$A \to B \to C \to D \to A \to E \to F \to G \to A$走刀路线，径向留余量0.1(直径值)，走刀路线如图4-4(b)，使用G90指令。

Step 3：Finish turn OD contour. The route is $A \to J \to K \to L \to M \to N \to P \to Q \to A$. The cutting route is shown in Figure 4-4（c）.

工步3：从右至左精车整个外圆表面及倒角。以$A \to J \to K \to L \to M \to N \to P \to Q \to A$的路线精车阶梯轴外圆和倒角，走刀路线如图4-4(c)所示。

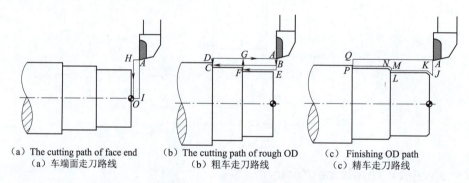

(a) The cutting path of face end
(a) 车端面走刀路线

(b) The cutting path of rough OD
(b) 粗车走刀路线

(c) Finishing OD path
(c) 精车走刀路线

Figure 4-4　The Cutting Path

图 4-4　数控加工路线

Step 4：cut off.

工步4：切断。

（4）Tool and Cutting Parameter

（4）选择刀具及切削参数

According to the surface quality requirements of the part，the material of the part and the material of the tool，select the cutting parameters by checking the cutting consumption table. The results are shown in Table 4-2.

根据零件的表面质量要求、零件的材料、刀具的材料等查切削用量表，刀具的切削参数，结果见表4-2。

(5) CNC Machining Process Card
（5）数控加工工序卡

CNC machining process card is shown in Table 4-2.

阶梯轴数控加工工序卡见表4-2。

Table 4-2 CNC Machining Process Card

CNC machining process card			Product name or code	Part Name	material	Part drawing number	
			002	Stepped shaft	45#	A4	
Process number	Program number	Chuck name	Chuck number	Equipment		workshop	
03	0202	Three-jaw chuck	01	CNC lathes		engineering shop	
Step number	Operation		Tool	Tool description	speed (r/min)	speed (mm/r)	Depth of Cut (mm)	Note
1	Face end		T01	Turning tool (cutting edge angle is 90°)	600	0.1		
2	Rough turn OD		T01	Turning tool (cutting edge angle is 90°)	600	0.2	< Radial 2	
3	Finish turn OD		T01	Turning tool (cutting edge angle is 90°)	1 000	0.1	Radial 0.5	
4	Cut off		T02	3mm Parting tool	500	0.05		

表 4-2 阶梯轴数控加工工序卡

数控加工工序卡			产品名称或代号	零件名称	材料	零件图号	
				阶梯轴	45#		
工序号	程序编号	夹具名称	夹具编号	使用设备		车间	
03	0202	三爪卡盘	01	数控车		机加工车间	
工步号	工步内容		刀具	刀具规格	主轴转速 (r/min)	进给量 (mm/r)	背吃刀量 (mm)	备注
1	车端面		T01	主偏角90°外圆车刀	600	0.1		
2	外圆粗加工		T01	主偏角90°外圆车刀	600	0.2	<径向2	
3	外圆精加工		T01	主偏角90°外圆车刀	1 000	0.1	径向0.5	
4	切断加工		T02	3 mm外切槽刀	500	0.05		

2. Program

2. 数控加工程序

The program origin is placed at the center of the right face of the workpiece. The start point of G94 and G90 is A (X42, Z2). This point is usually closer to the workpiece, but there is a certain distance from the blank. The tool change point is set at (X100, Z100). The radial dimension tolerance of the OD is large, therefore, the programming is based on the average size. Replace $\phi34$ with $\phi33.95$ and replace $\phi38$ with $\phi37.95$ in finish OD contour. The CNC machining program is shown in Table 4-3.

工件坐标系原点放在工件右端面中心，车端面、车外圆循环起点为A，坐标定为(X42, Z2)，该点通常离工件较近，但与毛坯留有一定的距离。换刀点设在(X100, Z100)处。外圆径向尺寸公差较大，所以按平均尺寸来编程，在精车时轴径 $\phi34$ 按 $\phi33.95$ 来编程，轴径 $\phi38$ 按 $\phi37.95$ 来编程。数控加工程序单见表4-3。

Table 4-3 The Programming and Meaning

Part Number	002	Part Name	Stepped shaft	The program origin prepared	Right face end center
Program number	00202	CNC system	FANUC 0iT		
Block number	World address command		Meaning		
N10	T0101;		Index to tool 1, use offset values in offset file 1		
N20	G97 G99M3 S600;		Cancel constant surface speed control, spindle on (CW) at 600 r/min		
N30	G0 X42 Z2;		Rapid move to start point O (42.2)		
N40	G94 X0 Z0 F0.1;		Face cycle G94		
N50	G90 X38.5 Z-35.8 F0.2;		Turning cycle G90		
N60	X34.5 Z-15.8;		Turning cycle G90		
N70	S1000M3;		Cancel constant surface speed control, spindle on (CW) at 1 000 r/min		
N80	G0 X28;		Rapid tool to J		
N90	G1 X33.95 Z-1;		Cut to K at feed rate		
N100	G1 Z-16;		Cut to L		
N110	X36;		Cut to M		
N120	X37.95 W-1;		Cut to N		
N130	Z-40;		Cut to P		
N140	X42;		Move to Q		
N150	G0 X100;		Quick retreat the tool change point		
N160	Z100;				
N170	M5;		Spindle off		
N180	M30;		Program end Memory reset		

表 4-3 数控加工程序单

零件号		零件名称	阶梯轴	编程原点	右端面中心
程序号	O0202	数控系统	FANUC 0iT	编制	
程序段号	程序内容		程序说明		
N10	T0101;		选择1号刀,刀补号为1号		
N20	G97 G99M3 S600;		主轴正转,设置粗车转速为600 r/min		
N30	G0 X42 Z2;		刀具快速定位到起始点O(42,2)		
N40	G94 X0 Z0 F0.1;		车端面G94循环		
N50	G90 X38.5 Z-35.8 F0.2;		车外圆G90循环		
N60	X34.5 Z-15.8;		车外圆G90循环		
N70	S1000M3;		设定精车转速1 000 r/min		
N80	G0 X28;		快速进刀到J点		
N90	G1 X33.95 Z-1;		使用进给量切到K		
N100	G1 Z-16;		切到L		
N110	X36;		切到M		
N120	X37.95 W-1;		切到N		
N130	Z-40;		切到P		
N140	X42;		快速到Q		
N150	G0 X100;		快速回到换刀点		
N160	Z100;				
N170	M5;		主轴停止		
N180	M30;		程序结束,重置内存		

4.3　Project 3　Program a Complex Stepped Shaft

4.3　项目3　复杂阶梯轴数控编程

4.3.1　Project Import

4.3.1　项目导入

The shape of the workpiece is shown in Figure 4-5. The material is 45 steel, and the blank is ϕ45 bar. Formulate the machining process of the part, and write the rough and finishing turning program for the part.

工件形状如图4-5所示,材料为45#钢,毛坯为 ϕ45 棒料,试制订零件的加工工艺,编写该零件的粗、精车程序。

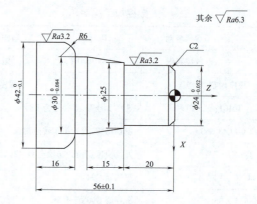

Figure 4-5　Case 4.3

图 4-5　例 4.3

The part is composed of straight lines and arcs, and the machining allowance in the X direction is relatively large. A G90 word executes repetitive straight-cut machining passes required to turn a part form stock. For complex-shaped surfaces with a large radial allowance, G71 word is used to complete the rough turning of the part. G71 and G70 words are used to complete the analysis of the machining process and CNC program.

该零件表面由直线和圆弧组成，径向的加工余量比较大。由于 G90 循环完成直线切削。对于复杂形状表面，且径向余量比较大，用 G71 指令完成零件的粗加工。本任务使用 G71、G70 指令完成含复杂阶梯轴的加工工艺分析和数控加工程序的编制。

4.3.2　Project Implementation

4.3.2　项目实施

1. Technology Analysis

1. 零件加工工艺分析

（1）Part Drawing

（1）零件图

The part in Figure 4-5 gradually increases in diameter from right to left. Therefore, it is possible to consider a process method which the entire one-time installation is used. $\phi 42$, $\phi 30$ and $\phi 24$ are critical dimensions, and the surface roughness is Ra 3.2. The rough and finish turning on the CNC lathe can meet the requirements.

图 4-5 中的零件从右端到左端直径逐渐变大。因此，可以考虑整体一次性安装加工完成的工艺方法。该零件的重要加工部位主要是 $\phi 42$、$\phi 30$ 和 $\phi 24$ 外圆，表面粗糙度为 Ra 3.2，在数控车床上分粗、精加工可以达到要求。

（2）Clamping Method

（2）装夹方案

The blank is a stock, which is positioned and clamped with a three-jaw chuck. The length of the workpiece protruding from the jaw is about 62 mm. The program origin is set at the right face end of the workpiece, and the tool change point is set ($X100, Z150$)

毛坯为棒料,用三爪自定心卡盘定位夹紧,工件伸出卡爪的长度约为 62 mm。工件零点设在工件右端面,换刀点设在工件右下方,Z 向距工件右端面 100 mm、X 向距轴心线 150 mm 的位置。

(3) Processing Sequence and Cutting Path

(3) 加工顺序及走刀路线

The workpiece is divided into rough and finish turning. The rough turning can use the G71 word. The finish turning uses G70 word to machine the part, and finally the workpiece is cut with a cutter.

工件分粗、精车进行加工,粗车可采用 G71 循环指令,精车用 G70 加工上述轮廓,最后用切断刀切断工件。

(4) Tool and Cutting Parameter

(4) 刀具及切削用量的选择

According to the surface quality requirements of the part, the material of the part, the material of the tool, select the cutting parameters by checking the cutting consumption table. The results are shown in Table 4-4.

根据零件的表面质量要求、零件的材料、刀具的材料等查切削用量表,选择计算刀具的切削参数,结果见表4-4。

(5) CNC Machining Processing Card

(5) 数控加工工序卡

Based on the analysis above, write CNC machining process card. See Table 4-4.

根据以上的分析,编写加工工序卡,见表4-4。

Table 4-4 CNC machining process card

CNC machining process card			Product name or code	Part Name	material	Part drawing number
			003	Handle	45#	A4
Process number	Program number	Chuck name	Chuck number	Equipment		workshop
03	0203	Three-jaw chuck	01	CNC lathes		Engineering shop
Step number	Operation		Tool	Tool description	Speed (r/min)	Feed (mm/r)	Depth of Cut(mm)	Note
1	Face end		T01	Turning tool (cutting edge angle is 90°)	800	0.1		
2	Rough turn OD		T01	Turning tool (cutting edge angle is 90°)		0.3	Radial 2	

续表

Step number	Operation	Tool	Tool description	Speed (r/min)	Feed (mm/r)	Depth of Cut(mm)	Note
3	Finish turn OD	T02	Turning tool (cutting edge angle is 93°)	1000	0.1	Radial 0.5	
4	Cut off	T03	Parting tool (the width of cutting edge is 3 mm)	600	0.05		

表 4-4 数控加工工序卡

数控加工工序卡		产品名称或代号		零件名称	材料		零件图号	
		003		手柄	45#钢		A4	
工序号	程序编号	夹具名称	夹具编号	使用设备			车间	
03	0203	三爪卡盘	01	数控车			机加工车间	
工步号	工步内容		刀具	刀具规格	主轴转速 (r/min)	进给量 (mm/r)	背吃刀量 (mm)	备注
1	车端面		T01	主偏角 90°外圆车刀	800	0.1		
2	外圆粗加工		T01	主偏角 90°外圆车刀	800	0.3	径向 2	
3	外圆精加工		T02	主偏角 93°外圆车刀	1 000	0.1	径向 0.5	
4	切断加工		T03	主偏角 3 mm 宽外割刀	600	0.05		

2. Program

2. 数控加工程序

CNC machining program is shown in Table 4-5.

数控加工程序单见表 4-5。

Table 4-5 The Programming and Meaning

Part Number	003	Part Name	handle	The program origin	Center of right face
Program number	O0203	CNC system	FANUC 0iT	prepared	
Block number	World address command		Meaning		
N10	T0101;		Index to tool 1, use offset values in offset file 1.		
N20	G97 G99 M3 S800;		Cancel constant surface speed control, spindle on (CW) at 800 r/min		
N30	G0 X46 Z2;		Rapid move to start point(46.2)		
N40	G94 X0 Z0 F0.3;		Face cycle G94		
N50	G71 U2 R0.5;		Rough turn contour defined by blocks: N70-N150, leave 0.5 mm along the X-axis, cut to 2 mm per side for each roughing pass, retreat distance is 0.5 mm		
N60	G71 P70 Q150 U1 W0 F0.3;				

续表

Part Number	003	Part Name	handle	The program origin	Center of right face
Program number	O0203	CNC system	FANUC 0iT	prepared	
Block number	World address command		Meaning		
N70	G0 X16;		Rapid move to (16,0)		
N80	G1 X24 Z-2 F0.1;		Cut to (20, -2)		
N90	Z-20;		Cut to (20, -20)		Finish turn contour
N100	X25;		Cut to (25, -20)		
N110	X30 Z-35;		Cut to (30, -35)		
N120	Z-40;		Cut to (30, -40)		
N130	G3 X42 W-6 R6;		Cut R6 to (42, -46)		
N140	G1 Z-60;		Cut to (42, -60)		
N150	X46;		Rapid move to (46, -60)		
N160	G0 X100;		Rapid move to the tool change point		
N170	Z100;				
N180	T0202;		Index to tool 2, use offset values in offset file 2		
N200	G97 G99 M3 S1000;		Cancel constant surface speed control, spindle on (CW) at 1000rpm		
N210	G0 X46 Z2;		Rapid move to start point(46.2)		
N220	G70 P70 Q150;		Finishing cycle G70		
N225	G0 X100;		Rapid move to the tool change point		
N230	Z100;				
N240	T0303;		Index to tool 3, use offset values in offset file 3.		
N250	G97 G99 M3 S600;		Cancel constant surface speed control, spindle on (CW) at 600 r/min		
N260	G0 Z-59;		Rapid move to cut-off position		
N270	X46;		Rapid move to (46, -59)		
N280	G1 X0 F0.05;		Cut off		
N290	G0 X100;		Rapid move to the tool change point		
N300	Z100;				
N310	M5;		Spindle off		
N320	M30;		Program end. Memory reset		

表4-5 数控加工程序单

零件号	003	零件名称	手柄	编程原点	右端面中心
程序号	O0203	数控系统	FANUC 0iT	编制	
程序段号	程序内容		程序说明		
N10	T0101;		换选择1号刀,刀具补偿号为1号		
N20	G97 G99 M3 S800;		主轴正转,设定主轴转速为800 r/min		
N30	G0 X46 Z2;		刀具快速定位到循环起点(46.2)		
N40	G94 X0 Z0 F0.3;		车右端面		

177

续表

零件号	003	零件名称	手柄	编程原点	右端面中心
程序号	O0203	数控系统	FANUC 0iT	编制	
程序段号	程序内容		程序说明		
N50	G71 U2 R0.5;		粗车循环,背吃刀量为 2 mm,退刀量为 0.5 mm;X 向精加工余量 0.5 mm,Z 向 0 mm		
N60	G71 P70 Q150 U1 W0 F0.3;				
N70	G0 X16;		快速进刀到(16,0)		精加工轮廓程序段
N80	G1 X24 Z-2 F0.1;		切到(20,-2)		
N90	Z-20;		切到(20,-20)		
N100	X25;		切到(25,-20)		
N110	X30 Z-35;		切到(30,-35)		
N120	Z-40;		切到(30,-40)		
N130	G3 X42 W-6 R6;		切 R6 到 (42,-46)		
N140	G1 Z-60;		切到(42,-60)		
N150	X46;		快速退刀(46,-60)		
N160	G0 X100;		快速回到换刀点		
N170	Z100;				
N180	T0202;		换 2 号刀,刀具补偿号为 2 号		
N200	G97 G99 M3 S1000;		设主轴转速为 1 000 r/min		
N210	G0 X46 Z2;		刀具快速定位到工件附近		
N220	G70 P70 Q150;		精加工循环		
N225	G0 X100;		快速回到换刀点		
N230	Z100;				
N240	T0303;		换 3 号刀,刀具补偿号为 3 号		
N250	G97 G99 M3 S600;		设主轴转速为 600 r/min		
N260	G0 Z-59;		快速定位到切断位置		
N270	X46;		快速进刀到(46,-59)		
N280	G1 X0 F0.05;		切断		
N290	G0 X100;		快速回到换刀点		
N300	Z100;				
N310	M5;		主轴停止		
N320	M30;		程序结束,内存重置		

4.4 Project 4 Program a Shaft with Arcs

4.4 项目4 带圆弧轴的数控编程

4.4.1 Project Import
4.4.1 项目导入

The shape of the workpiece is shown in Figure 4-6. The material is 45 steel, and the blank is ϕ25 bar. Try to write the machining process of the part and write the rough and finish turning program for the part.

工件形状如图4-6所示,材料为45#钢,毛坯为ϕ25棒料,试制订零件的加工工艺,编写该零件的粗、精车程序。

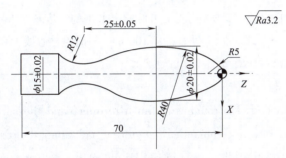

Figure 4-6　Case 4.4
图4-6　例4.4

The surface of the part is composed of straight lines and arcs, and the machining allowance in the radial direction is relatively large. The G90 cycle can only solve the cutting problem of cylindrical and tapered surfaces, and the cutting depth must be calculated every time. For complex-shaped surfaces with a large radial allowance, the compound cycle instruction G73 is used to complete the rough and finish turning.

该零件表面由直线和圆弧组成,径向的加工余量比较大。由于G90循环只能解决圆柱面和锥面切削问题,且要计算每次的切削深度。对于复杂形状表面,且径向余量比较大,用G73指令完成零件的粗精加工。

4.4.2 Project Implementation
4.4.2 项目实施

1. Processing Technology Analysis

1. 加工工艺分析

（1）Part Drawing Analysis

（1）零件图分析

The left of the part in Figure 4-6 has a small outer diameter, and the right is a convex and concave arcs surface. Therefore, it is possible to consider a process method which one-time installation is used. $\phi15$ and $\phi20$ are critical dimensions, and the surface roughness is Ra 3.2. The rough and finish turning can meet the requirements.

图 4-6 中零件的左端为直径较小的外圆,右端为凸、凹圆弧面。因此,可以考虑整体一次性安装加工完成的工艺方法。该零件的重要加工部位主要是 $\phi15$ 及 $\phi20$ 外圆,表面粗糙度为 Ra 3.2,在数控车床上分粗、精加工可以达到要求。

(2) Clamping Method
(2) 装夹的选择

The blank is a bar, and it is positioned and clamped with a three-jaw chuck. The program origin is set at the right end face of the workpiece, and the tool change point is set at the right side of the workpiece. The Z direction is 100 mm from the right face of the workpiece, and the X direction is 150 mm.

毛坯为棒料,用三爪自定心卡盘定位夹紧,工件伸出卡爪的长度约为 80 mm。工件零点设在工件右端面,换刀点设在工件方,Z 向距工件右端面 100 mm、X 向距轴心线 150 mm 的位置。

(3) Processing Sequence and Cutting Path
(3) 确定加工顺序及走刀路线

The workpiece is divided into rough and finish turning, and the surface of this part has a concave arc. Since the G73 cycle commanded the cutting path along the direction parallel to the Z axis to cut from the right to the left, it could not be cut at the concave arc during rough machining. Use G73 command to cycle the arc contour and $\phi16$ outer circle. Finish the contour with G70, and finally cut the workpiece with a cutter.

工件分粗、精车进行加工,该零件表面有凹圆弧,由于 G71 循环指令走刀路线沿着平行于 Z 轴的方向从右往左分层切削,在粗加工时内凹圆弧处没法切削,用 G73 指令循环加工圆弧轮廓和 $\phi16$ 外圆,精车用 G70 加工上述轮廓,最后用切断刀切断工件。

(4) Tool and Cutting Parameter
(4) 刀具及切削用量的选择

According to the machining accuracy requirements of the part, the material of the part and tool, consult the relevant cutting manual. The values are shown in Table 4-6.

根据零件的加工精度要求、零件的材料、刀具的材料查相关切削手册,具体数值见表 4-6。

(5) CNC Machining Process Card
(5) 填写数控加工工序卡

CNC machining process card is shown in Table 4-6.

根据以上的分析,编写数控加工工序卡,见表 4-6。

Table 4-6 CNC Machining Process Card

CNC machining process card		Product name or code	Part Name	material	Part drawing number
		004	handle	45#	A4
Process number	Program number	Chuck name	Chuck number	Equipment	workshop
003	0204	Three-jaw chuck	01	CNC lathes	Engineering shop
Step number	Operation		Tool	Tool description	Spindle speed (m/min)	Feed speed (mm/r)	Depth of Cut (mm)	Note
1	Face end		T01	Turning tool (cutting edge angle is 90°)	120	0.1		
2	Rough turn OD		T01	Turning tool (cutting edge angle is 90°)		0.2	Radial 1.5	
3	Finish turn OD		T02	Turning tool (cutting edge angle is 93°)	150	0.1	Radial 0.5	
4	Cut off		T03	Parting tool (the width of cutting edge is 3 mm)	80	0.05		

表 4-6 数控加工工序卡

数控加工工序卡		产品名称或代号	零件名称	材料	零件图号
		004	手柄	45钢	A4
工序号	程序编号	夹具名称	夹具编号	使用设备	车间
003	0204	三爪卡盘	01	数控车	机加工车间
工步号	工步内容		刀具	刀具规格 (mm)	主轴转速 (m/min)	进给量 (mm/r)	背吃刀量 (mm)	备注
1	车端面		T01	主偏角90° 外圆车刀	120	0.1		
2	外圆粗加工		T01	主偏角90° 外圆车刀		0.2	径向1.5	
3	外圆精加工		T02	主偏角93° 外圆车刀	150	0.1	径向0.5	
4	切断加工		T03	3 mm 宽外割刀	80	0.05		

2. Mathematical Process

2.数学处理

The program origin is determined at the center of the right end face of the workpiece. The coordinates of points A, B, and C are obtained by using CAD drawing software: $A(8.572, -2.425)$; $B(10.148, -42.259)$; $C(15, -56.885)$, as shown in Figure 4-7.

编程坐标系原点定于工件右端面中心,在 CAD 绘图软件中通过作图,求得点 A、B、C 坐标分别为:$A(8.572, -2.425)$;$B(10.148, -42.259)$;$C(15, -56.885)$,如图 4-7 所示。

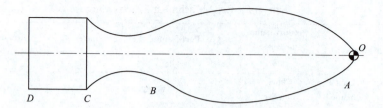

Figure 4-7 Calculate the base point coordinates

图 4-7 基点坐标的计算

3. Program

3. 数控加工程序

The CNC machining program list is shown in Table 4-7.

数控加工程序单见表 4-7。

Table 4-7 The Programmings and Meanings

Part Number	004	Part Name	handle	The program origin prepared	Center of right face
Program number	O0202	CNC system	FANUC 0iT		
Block number	World address command		Meaning		
N10	T0101;		Index to tool 1, use offset values in offset file 1.		
N20	G96 G99 M3 S120;		Constant surface speed control, spindle on (CW) at 120 m/min		
N30	G50 S2500;		Set maximum spindle speed at 2 500 r/min		
N40	G0 X40 Z2;		Rapid move to the start point (40,2)		
N50	G94 X0 Z0 F0.1;		Face end		
N60	G73 U11 W0 R8;		Rough cycle G73, rough contour defined by blocks 80–150. Leave 0.5 mm Per side alone the X-axis. Cut at 0.2 Feed.		
N70	G73 P80 Q150 U1 W0 F0.2;				
N80	G0 X-1 Z2;		Rapid to (-1,2)		
N90	G1 Z0 F0.1;		Cut to (-1,0)		
N100	X0;		Cut to O		Finish turn blocks
N110	G3 X8.572 Z-2.425 R5;		Cut to A		
N120	X10.148 Z-42.259 R40;		Cut to B		
N130	G2 X15 Z-56.885 R12;		Cut to C		
N140	G1 Z-75;		Cut to D		
N150	X18;		Cut to (18, -75)		
N160	X100;		Rapid move to the tool change point		
N170	Z150;				
N180	T0202;		Index to tool 2, use offset values in offset file 2		
N200	G96 S150 M3;		Constant surface speed control, spindle on (CW) at 150 mpm		
N210	G0 X40 Z2;		Rapid move to the start point (42,2)		
N220	G50 S2500;		Set maximum spindle speed at 2 500 r/min		
N225	G70 P80 Q150;		Finish turning cycle		

续表

Part Number	004	Part Name	handle	The program origin	Center of right face
Program number	O0202	CNC system	FANUC 0iT	prepared	
Block number	World address command			Meaning	
N230	X100；			Rapid move to the tool change point	
N240	Z150；				
N250	T0303			Index to tool 3, use offset values in offset file 3	
N260	G96 S80 M3；			Constant surface speed control, spindle on (CW) at 80 m/min	
N270	G50 S2500；			Set spindle maximum speed at 2 500 r/min	
N280	G0 Z-73；			Rapid move to cut-off position	
N290	X20；				
N300	G1 X0 F0.05；			Cut off	
N310	X100；			Rapid move to the tool change point	
N320	Z150；				
N330	M5；			Spindle off	
N340	M30；			Program end. Memory reset	

表 4-7 数控加工程序

零件号	004	零件名称	手柄	编程原点	右端面中心
程序号	O0204	数控系统	FANUC 0iT	编制	
程序段号	程序内容			程序说明	
N10	T0101；			换选择1号刀,刀具补偿号为1号	
N20	G96 G99 M3 S120；			主轴正转,设定主轴转速为120 m/min	
N30	G50 S2500；			设定主轴最高转速2 500 r/min	
N40	G0 X40 Z2；			刀具快速定位到循环起点(40,2)	
N50	G94 X0 Z0 F0.1；			车右端面	
N60	G73 U11 W0 R8；			粗车,精车轮廓程序段80-150,精车X向留0.5 mm,进给量0.2 mm/r	
N70	G73 P80 Q150 U1 W0 F0.2；				
N80	G0 X-1 Z2；			快速进刀到(-1,2).	
N90	G1 Z0 F0.1；			切到(-1,0).	
N100	X0；			切到O点	
N110	G3 X8.572 Z-2.425 R5；			切到A点	精加工轮廓程序段
N120	X10.148 Z-42.259 R40；			切到B点	
N130	G2 X15 Z-56.885 R12；			切到C点	
N140	G1 Z-75；			切到D点	
N150	X18；			切到(18,-75)	
N160	X100；			快速至换刀点	
N170	Z150；				
N180	T0202；			换2号刀,刀具补偿号为2号	
N200	G96 S150 M3；			主轴正转,设定主轴转速为150 m/min	
N210	G0 X40 Z2；			刀具快速定位到循环起点(40,2)	

续表

零件号	004	零件名称	手柄	编程原点	右端面中心
程序号	O0204	数控系统	FANUC 0iT	编制	
程序段号	程序内容		程序说明		
N220	G50 S2500;		设定主轴最高转速 2 500 r/min		
N225	G70 P80 Q150;		精加工循环		
N230	X100;		快速退刀至换刀点		
N240	Z150;				
N250	T0303		换 3 号刀,刀具补偿号为 3 号		
N260	G96 S80 M3;		主轴正转,设定主轴转速为 80 m/min		
N270	G50 S2500;		设置主轴最高转速 2 500 r/min		
N280	G0 Z-73;		快速定位到切断位置		
N290	X20;		X 方向快速进刀		
N300	G1 X0 F0.05;		切断		
N310	X100;		快速回到换刀点		
N320	Z150;				
N330	M5;		主轴停止		
N340	M30;		程序结束,内存重置		

4.5　Project 5　Program a Threaded Shaft

4.5　项目 5 螺纹轴的数控编程

4.5.1　Project Import

4.5.1　项目导入

As shown in Figure 4-8, the blank is $\phi40$ bar stock and the material is 45 steel. Analyze the NC machining process of the part, and write the NC machining program.

如图 4-8 所示螺栓,毛坯为 $\phi40$ 棒料,材料为 45#钢。分析零件的数控加工工艺,并编制数控加工程序。

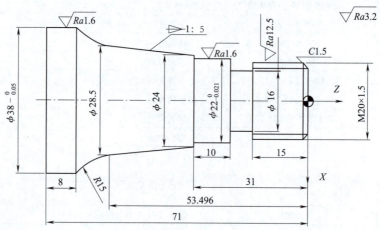

Figure 4-8　Case 4.5

图 4-8　例 4.5

4.5.2 Project Implementation
4.5.2 项目实施

1. Processing Technology
1. 零件加工工艺分析

(1) Part Drawing
(1) 零件图分析

As shown in Figure 4-8, the part consists of a cylindrical surface, a conical surface, a groove, and a thread. The critical dimensions are $\phi38$ and $\phi22$, and the surface roughness is Ra 1.6, the finish turning can meet the requirements. The surface roughness of $\phi16$ is Ra 12.5, the rough turning can meet the requirements.

如图 4-8 所示,零件由圆柱面、圆锥面、槽和螺纹组成,该零件的重要加工部位主要是 $\phi38$ 及 $\phi22$ 外圆,表面粗糙度为 Ra 1.6,在数控车床上精加工可以达到要求,$\phi16$ 的槽表面粗糙度为 Ra 12.5,在数控车床上粗加工可以达到要求。

(2) Clamping Method
(2) 装夹选择

Select universal clamp, three-jaw automatic centering chuck for clamping.

选择通用夹具,三爪自动定心卡盘夹紧。

(3) Processing Sequence
(3) 加工顺序的确定

Step 1: Turn the end face.

工步 1:车端面。

Step 2: Rough turn the OD surface.

工步 2:粗车外表面。

Step 3: Finish turn the OD surface.

工步 3:精车外表面。

Step 4: Groove.

工步 4:切退刀槽。

Step 5: Cut threads.

工步 5:车螺纹。

Step 6: Cut off.

工步 6:切断。

(4) Tool and Cutting Parameter
(4) 刀具和切削参数

According to the processing requirements of parts, turning tools, external thread turning tools and cutting tools are required. The same tool is used for rough and finish turning of the OD surface to avoid changing tools.

根据零件的加工要求,需要外圆车刀、外螺纹车刀和切断刀,为了避免换刀,外圆的粗精加工用同一把刀。

According to the machining accuracy requirements of the part, the material of the part and tool, consult the relevant cutting manual. The values are shown in Table 4-8.

根据零件的加工精度要求、零件的材料、刀具的材料查相关切削手册,具体数值见表4-8。

(5) CNC Machining Process Card

(5)数控加工工序卡

Table 4-8 shows the NC turning process cards for bolt parts.

螺栓零件的数控加工工序卡见表4-8。

Table 4-8 CNC Machining Process Card

CNC machining process card			Product name or code	Part Name	material	Part drawing number	
			005	bolt	45#	A4	
Process number	Program number	Chuck name	Chuck number	Equipment		workshop	
03	0204	Three-jaw chuck	01	CNC lathes		Engineering shop	
Step number	Operation		Tool	Tool description	Spindle speed (r/min)	Feed speed (mm/r)	Depth of Cut (mm)	Note
1	Face end		T01	Turning tool (cutting edge angle is 93°)	600	0.1		
2	Rough turn OD		T01		600	0.3	Radial 2	
3	Finish turn OD		T01		1 000	0.1	Radial 0.5	
4	Groove		T03	Parting tool (the width of cutting edge is 4 mm)	300	0.05		
5	Cut threads		T04	OD threading tool	400			
6	Cut off		T03	Parting tool (the width of cutting edge is 3 mm)	300	0.05		

表 4-8　数控加工工序卡

数控加工工序卡	产品名称或代号		零件名称		材料		零件图号	
	005		螺栓		45钢		A4	
工序号	程序编号	夹具名称	夹具编号	使用设备			车间	
03	0205	三爪卡盘	01	数控车			机加工车间	
工步号	工步内容		刀具	刀具规格（mm）	主轴转速（r/min）	进给量（mm/r）	背吃刀量（mm）	备注
1	车端面		T01	主偏角93°外圆车刀	600	0.1		
2	粗车外圆		T01		600	0.3	径向2	
3	精车外圆		T01		1000	0.1	径向0.5	
4	切退刀槽		T03	切槽刀（宽4 mm）	300	0.05		
5	车螺纹		T04	螺纹车刀	400			
6	切断加工		T03	切槽刀（宽3 mm）	300	0.05		

2. Program

2. 数控加工程序

Set the program origin at the center of the right face end and the tool change point at ($X100$, $Z100$). The dimension of the shaft diameter before thread processing is: $d_{圆柱} \approx 20 - 0.2 = 19.8$, thread depth is $h \approx 1.299$, small thread diameter is: $d \approx 17.4$. The Programmings and Meanings see Table 4-9.

将工件坐标原点设在右端面的中心，换刀点设为($X100$, $Z100$)。螺纹加工前轴径的尺寸为：$d_{圆柱} \approx 20 - 0.2 = 19.8$，牙深为$h \approx 1.299$，螺纹小径$d \approx 17.4$。数控加工程序见表4-9。

Table 4-9　The Programmings and Meanings

Part Number	005	Part Name	bolt	The program origin	Center of right face end
Program number	O0205	CNC system	FANUC 0iT	prepared	
Block number	World address command			Meaning	
N10	T0101;			Index to tool 1, use offset values in offset file number is No. 1	
N20	G97 G99 S600 M3;			Cancel constant surface speed control, spindle on (CW) at 600 r/min	
N30	G0 X42 Z2;			Rapid move to the start point(42.2)	
N40	G94 X0 Z0 F0.1			Machine face end	
N50	G71 U2 R1;			Rough turning cycle G71, rough contour defined by blocks 70 – 160. Leave 0.5 mm Per side alone the X-axis. Cut at 5 Feed	
N60	G71 P70 Q160 U1 W0 F0.3;				

续表

Part Number	005	Part Name	bolt	The program origin	Center of right face end
Program number	O0205	CNC system	FANUC 0iT	prepared	
Block number	World address command			Meaning	
N70	G0 X12.8;			Rapid move along X-axis	Finish turn blocks
N80	G1 X19.8 Z-1.5 F0.1			Cut chamfer	
N90	Z-21;			Cut OD ϕ19.8, finish tool position at Z-21	
N100	X22;			Cut step	
Z110	Z-31;			Cut OD ϕ22, finish tool position at Z-31	
N120	X24;			Cut step	
N130	X28.5 Z-53.496;			Cut OD, finish tool at X 28.5, Z-53.496	
N140	G2 X38 Z-63 R15;			Cut arc R15	
N150	Z-74			Cut OD ϕ38, finish tool position at Z-74	
N160	X42;			Rapid move to (42, -74)	
N170	S1000 M3;			Cancel constant surface speed control, spindle on (CW) at 1 000 r/min	
N180	G70 P70 Q160;			Finish turning cycle G70	
N190	G0 X100;			Rapid move to the tool change point	
N200	Z100;				
N210	T0303;			Index to tool 3, use offset values in offset file number is No. 3	
N220	G97 G99 S300 M3;			Cancel constant surface speed control, spindle on (CW) at 300 r/min	
N230	G0 Z-19;			Rapid move to grooving position (25, -19)	
N240	X25;				
N250	G1 X16 F0.05			Groove	
N260	G04 X0.5;			Pause for 0.5 s	
N270	X25			Move along X-axis at feedrate	
N280	Z-21			move along Z-axis at feedrate	
N290	X16			Groove	
N300	G04 X0.5;			Pause for 0.5 s	
N310	G0 X100;			Rapid move to the tool change point	
N320	Z100;				
N330	T0404;			Index to tool 4, use offset values in offset file number is No. 4	
N340	G97 G99 S400 M3;			Cancel constant surface speed control, spindle on (CW) at 400 r/min	
N350	G0 X25 Z5;			Rapid tool to the start point (25.5)	
N360	G92 X19 Z-26.5 F2;			Cut, threads pass 1	
N370	X18.4;			pass 2	
N380	X18;			pass 3	
N390	X17.84;			pass 4	
N400	X17.84;			pass 4	
N410	G0 X100;			Rapid move to the tool change point	
N420	Z100;				

续表

Part Number	005	Part Name	bolt	The program origin	Center of right face end
Program number	O0205	CNC system	FANUC 0iT	prepared	
Block number	World address command			Meaning	
N430	T0303;			Index to tool 3, use offset values in offset file number is No. 3	
N440	G97 G99 S300 M3;			Cancel constant surface speed control, spindle on (CW) at 300 r/min	
N450	G0 Z-75;			Rapid move to cut-off position(42, -75)	
N460	X42;				
N470	G1 X0 F0.05;			Cut off	
N480	G0 X100;			Rapid move to the tool change point	
N490	Z100;				
N500	M5;			Spindle off	
N510	M30;			Program end. Memory reset	

表 4-9 数控加工程序

零件号	005	零件名称	螺栓	编程原点	右端面中心
程序号	O0205	数控系统	FANUC 0iT	编制	
程序段号	程序内容			程序说明	
N10	T0101;			换外圆车刀	
N20	G97 G99 S600 M3;			主轴正转,转速为 600 r/min	
N30	G0 X42 Z2;			快速定位到循环起始点	
N40	G94 X0 Z0 F0.1			车端面	
N50	G71 U2 R1;			外圆粗车循环,精车轮廓程序段 70-160,精车 X 向留 0.5 mm, 进给量 3 mm/r	
N60	G71 P70 Q160 U1 W0 F0.3;				
N70	G0 X12.8;			X 向进刀	
N80	G1 X19.8 Z-1.5 F0.1			车倒角	
N90	Z-21;			车螺纹的外圆柱面 ϕ19 到 Z-21	
N100	X22;			车台阶	
Z110	Z-31;			车 ϕ22 外圆到 Z-31	外圆精车轮廓程序段
N120	X24;			车台阶	
N130	X28.5 Z-53.496;			车圆锥面	
N140	G2 X38 Z-63 R15;			车 R15 圆弧	
N150	Z-74;			车 ϕ38 外圆到 Z-74	
N160	X42;			退刀	
N170	S1000 M3;			主轴正转,转速为 1 000 r/min	
N180	G70 P70 Q160;			外圆精车循环	
N190	G0 X100;			回到换刀点	
N200	Z100;				
N210	T0303;			换切槽刀	
N220	G97 G99 S300 M3;			主轴正转,转速为 300 r/min	

189

续表

零件号	005	零件名称	螺栓	编程原点	右端面中心
程序号	O0205	数控系统	FANUC 0iT	编制	
程序段号	程序内容		程序说明		
N230	G0 Z-19;		快速定位到切槽位置(25,-19)		
N240	X25;				
N250	G1 X16 F0.05;		切槽		
N260	G04 X0.5;		槽底暂停0.5 s		
N270	X25;		按进给量X方向退刀		
N280	Z-21;		按进给量Z方向进刀		
N290	X16;		切槽		
N300	G04 X0.5;		暂停0.5 s		
N310	G0 X100;		回到换刀点		
N320	Z100;				
N330	T0404;		换螺纹刀		
N340	G97 G99 S400 M3;		主轴正转,转速为400 r/min		
N350	G0 X25 Z5;		刀具快速定位到循环起点(25,5)		
N360	G92 X19 Z-26.5 F2;		车削螺纹第一线程		
N370	X18.4;		车削螺纹第二线程		
N380	X18;		车削螺纹第三线程		
N390	X17.84;		车削螺纹第四线程		
N400	X17.84;		螺纹光整加工		
N410	G0 X100;		回到换刀点		
N420	Z100;				
N430	T0303;		换切槽刀		
N440	G97 G99 S300 M3;		主轴正转,转速为300 r/min		
N450	G0 Z-75;		快速定位到切断位置(42,-75)		
N460	X42;				
N470	G1 X0 F0.05;		切断		
N480	G0 X100;		回到换刀点		
N490	Z100;				
N500	M5;		主轴停止		
N510	M30;		程序结束,重置内存		

Appendix A　Important Safety Precautions

附录A　重要安全措施

The reader is strongly advised to study and understand all safety precautions before entering the shop area. Special attention should be given to applying these precautions when working in the shop.

建议读者在进入车间区域之前学习并了解所有安全预防措施。在车间内工作时,应特别注意采取这些预防措施。

1. Personal Attire and Personal Safeguards

(1) Wear ANSI-approved safety goggles and a protective shop apron.

(2) Avoid loose clothing and accessories (neckties, gloves, watches, rings, etc.).

(3) Put long hair up under an approved shop cap.

(4) Avoid skin contact with cutting fluids.

(5) For cutting operation in excess of OSHA limits, wear a face mask.

(6) Flat, nonslip safety shoes are recommended.

(7) Wear hearing protection for noise levels above OSHA specifications.

(8) Wear a face mask for dust levels above OSHA limits.

(9) Do not operate a CNC machine while under the influence of drugs (prescribed or otherwise).

(10) Regardless of how slight the injury may be, always notify the instructor. Apply first-aid treatment to any cuts or bruises.

1. 个人穿戴和个人安全指南

(1) 佩戴 ANSI 认可的护目镜和防护服。

(2) 避免穿宽松的衣服和佩戴装饰品(领带、手套、手表、戒指等)。

(3) 将长发放在帽子里。

(4) 避免皮肤接触切削液。

(5) 对于超过 OSHA 限制的切削操作,请戴上口罩。

(6) 建议使用扁平防滑安全鞋。

(7) 当噪声超过 OSHA 规定时,佩戴听力保护装置。

(8) 粉尘高于 OSHA 限值时,佩戴面罩。

(9) 在药物(处方药或其他)的影响下,不要操作 CNC 机器。

（10）无论受伤有多轻，请务必通知讲师。对任何伤口或瘀伤进行急救治疗。

2. Shop Environment

（1）Keep the floor free from oil and grease.

（2）Remove chips from the floor. They can become embedded in the soles of shoes and cause dangerous slippage.

（3）Keep tools and materials off the floor.

2. 车间环境

（1）保持地板不含油和油脂。

（2）从地板上清除切屑。它们可能嵌入鞋底并导致危险。

（3）不要将工具和材料放在地上。

3. Tool Selection and Handling

（1）Store tools in the tool tray.

（2）Make sure tools are sharp and in good condition.

（3）Grind carbide or ceramic tools only in a well-ventilated area. Do not grind near any CNC machine tool.

（4）For carbide or ceramic insert tools, use the thickest insert possible.

（5）Select the thickest and shortest tool holder possible.

（6）Check all seating areas on holders to be sure each cutting tool rests solidly.

（7）Use only approved tools for the job.

（8）Do not exceed the manufacturers' recommended rpm（r/min）for the tool.

（9）Do not force any tool.

3. 工具选择和处理

（1）将刀具存放在刀架上。

（2）确保刀具锋利且状态良好。

（3）仅在通风良好的区域内研磨碳化物或陶瓷刀具。不要在任何数控机床附近磨削。

（4）对于硬质合金或陶瓷刀片，请尽可能使用最厚的刀片。

（5）选择最粗和最短的刀杆。

（6）检查刀架上的所有区域，确保每个刀具都牢牢固定。

（7）仅使用经批准的刀具进行工作。

（8）不要超过制造商推荐的刀具主轴转速（r/min）。

（9）不要强行使用任何工具。

4. CNC Machine Tool Handling

（1）Secure all adjusting keys and wrenches before machining.

（2）Check compressed air equipment and any damaged parts on the CNC machine.

（3）Check the oil levels.

(4) Use only recommended accessories.

(5) Be sure the CNC machine is connected to a grounded, permanent electrical power box.

(6) Keep the work area well lighted.

(7) Be mindful of obstructions and sharp cutting tools when learning into the work area.

4. 数控机床加工指南

(1) 在加工前确保所有键盘和扳手的安全。

(2) 保证压缩空气设备工作正常和确保数控机床上的没有损坏部件。

(3) 检查油位。

(4) 仅使用推荐的配件。

(5) 确保数控机床连接到接地的永久电源箱。

(6) 保持工作区域明亮。

(7) 在工作区时要注意障碍物和锋利的刀具。

5. *Machining Practices*

(1) In case of any emergency while operating a CNC machine, hit the EMERGENCY STOP button.

(2) Prior to its operation, be sure to check that there are no obstacles in the machine's working area.

(3) Check the position height and cross-travel movements of each tool to verify it will not collide with surroundings objects
- when the tool is being lowered to a machining surface;
- when the tool is executing a cutting operation;
- when the tool is being moved to the tool change position.

(4) Use manufacturers' tables of recommended cutting speeds and feeds. Adjust these parameters based on specifications of piece part accuracy quality of surface finish, rate of tool wean chip control, and machine capability.

(5) Make a dry run for safety checks.

(6) Check the workpiece to see if it is free of burrs and particles.

(7) Make a deep first cut below the hard outer scale on castings, forgings, and other rough or irregular surfaces.

(8) Supply a continuous flow of cutting fluid to carbide tools when machining cast iron or steel.

(9) Reduce the speed and feed when drilling large-diameter holes to a depth of more than twice the drill size.

(10) Use slower speeds for thread cutting than for other turning operations.

(11) Finish cut an internal taper by moving the tool in the direction of the larger diameter.

(12) Do not remove debris while the CNC machine is running.

(13) Always consult with the instructor before starting any unfamiliar operation.

5. 加工操作指南

(1) 如果在操作数控机床时发生紧急情况,请按下急停按钮。

(2) 在操作之前,请务必检查,确保机器工作区域内没有障碍物。

(3) 检查每把刀具的位置高度和交叉行程,以确认它不会与周围物体发生碰撞:

- 当工具降低到加工表面时;
- 当工具执行切削操作时;
- 当工具移动到换刀位置时。

(4) 使用制造商推荐的切削速度和进给量。根据工件表面粗糙度精度、刀具切削率、切削加工性能等指标调整这些参数。

(5) 使用空运行进行安全检查。

(6) 检查工件,看它是否有毛刺和颗粒。

(7) 在铸件、锻件和其他粗糙或不规则表面上的硬皮下方进行首次切削。

(8) 在加工铸铁或钢时,为硬质合金刀具提供切削液。

(9) 钻大直径孔的深度超过钻头尺寸的两倍时,降低切削速度和进给量。

(10) 与其他车削操作相比,螺纹的切削速度较慢。

(11) 通过沿较大直径方向移动刀具来完成内部锥度切削。

(12) 数控机床运行时不要清除切屑。

(13) 在开始任何非常规操作之前,请务必咨询老师。

Appendix B Summary of G Codes, M Codes and Auxiliary Functions for Turning Operations

附录B 数控车的G代码、M代码和辅助变量汇总

Table B-1 Summary of G Codes for Turning Operations (FANUC Controllers)

G code	Function	Mode
G0	Rapid positioning (traverse tool movement)	Modal
G1	Linear interpolation (tool movement at feed rate)	Modal
G2	Circular interpolation clockwise (CW)	Modal
G3	Circular interpolation counterclockwise (CCW)	Modal
G4	Programmed dwell (entered as a separate block)	Nonmodal
G9	Exact stop check (entered as a separate block)	Modal
G10	Data setting (programmed offset value setting) Tool nose radius compensation values for G41, G42 words	Modal
G11	Cancel data setting mode	Nonmodal
G20	Input data inches (in.)	Modal
G21	Input data in metric (mm)	Nonmodal
G22	Programmed safety zone (no tool entry)	Modal
G23	Cancel programmed safety zone	Modal
G25	Alarm when spindle speed deviates from program speed (OFF) Alarm	Modal
G26	when spindle speed deviates from program speed (ON)	Modal
G27	Alarm if tool does not reach, Reference point on return	Modal
G28	Return to reference point (machine zero)	Modal
G29	Return from reference point (machine zero)	Modal
G30	Return to second, third, and fourth reference point	Modal
G32	Constant lead thread cutting cycle (single pass)	Nonmodal
G34	Variable lead thread cutting cycle (single pass)	Modal
G40	Cancel tool nose radius compensation (TNR)	Modal
G41	Tool nose radius compensation left	Nonmodal
G42	Tool nose radius compensation right	Modal
G50	● Zero offset (programming of a temporary zero point) ● Maximum spindle speed setting for constant surface speed in rpm	Nonmodal
G65	Call user macro (one-time call)	Modal
G66	Call user macro (repeat call)	Modal
G67	Cancel G66 function	Modal
G68	Mirror image for double turrets (ON)	Modal
G69	Mirror image for double turrets (OFF)	Modal

续表

G code	Function	Mode
G70	Finish turning autocycle	Modal
G71	Rough turning autocycle	Modal
G73	Repeat a fixed pattern cut at a set of programmed location	Modal
G74	Peck drilling autocycle (2 P axis)	Nonmodal
G75	Groove autocycle (X-axis)	Modal
G76	Constant lead thread cutting cycle (advanced multiple pass)	Modal
G90	Turning/boring cutting cycle	Nonmodal
G92	Constant lead thread cutting cycle (basic multiple pass)	Modal
G94	End face turning cycle	Nonmodal
G96	Programmed constant surface speed control	Modal
G97	Cancel G96 command	Modal
G98	Feed rate inputted in mm/min	Modal
G99	Feed rate inputted in mm/r	Modal

表 B-1　用于车削操作的 G 代码汇总（FANUC 控制器）

G 指令	功能	模式
G0	快速定位（刀具移动运动）	模态
G1	线性插补（进给量下的刀具运动）	模态
G2	顺时针圆弧插补（CW）	模态
G3	逆时针圆弧插补（CCW）	模态
G4	程序停留（作为单独的程序段输入）	非模态
G9	停止检查（输入为单独的程序段）	模态
G10	数据设置（编辑 G41、G42 的刀尖圆弧半径补偿值）	模态
G11	取消数据设置模式	非模态
G20	输入英制数据（英寸）	模态
G21	输入公制数据（mm）	非模态
G22	存储行程检测功能有效（刀具不能进入）	模态
G23	存储行程检测功能无效	模态
G25	主轴速度偏离程序速度时报警	模态
G26	主轴转速偏离程序速度时不报警	模态
G27	如果刀具未到达则报警，返回参考点	模态
G28	返回参考点（机床原点）	模态
G29	从参考点返回（机床原点）	模态
G30	返回第二、第三和第四参考点	模态
G32	恒定的螺纹切削循环（单程）	非模态
G34	可变导程螺纹切削循环（单程）	模态
G40	取消刀尖圆弧半径补偿（TNR）	模态
G41	刀尖圆弧半径左补偿	非模态

Appendix B Summary of G Codes, M Codes and Auxiliary Functions for Turning Operations
附录B 数控车的G代码、M代码和辅助变量汇总

续表

G 指令	功能	模式
G42	刀尖圆弧半径右补偿	模态
G50	● 零点偏移(设定临时编程原点) ● 设置恒定线速度的最大主轴速度转速	非模态
G65	调用宏(一次调用)	模态
G66	调用宏(重复调用)	模态
G67	取消 G66 功能	模态
G68	双刀架的镜像车削(开)	模态
G69	双刀架的镜像车削(关)	模态
G70	精车循环	模态
G71	粗车循环	模态
G73	固定形状粗车循环	模态
G74	钻孔自动循环(Z 轴)	非模态
G75	切槽自动循环(X 向)	模态
G76	恒定的螺纹切削循环(高级多次线程加工)	模态
G90	车削/镗孔循环	非模态
G92	恒定的螺纹切削循环(多次车削)	模态
G94	端面车削循环	非模态
G96	程序恒定线速度控制	模态
G97	取消 G96 命令	模态
G98	按 mm/min 输入的进给量	模态
G99	按 mm/r 输入的进给量	模态

Table B-2 Summary of M Codes for Turning Operations (FANUC Controllers)

M code	Function
M0	program stop
M1	Optional program stop
M2	End of program
M3	Spindle on clockwise (CW)
M4	spindle on counterclockwise (CCW)
M5	Spindle off
M6	Program stop (automatic tool change)
M8	Coolant on
M9	Coolant off
M13	Spindle on clockwise/coolant on
M14	Spindle on counterclockwise/coolant on
M17	Spindle off/coolant off
M19	Oriented Spindle off
M30	Program end, memory reset

续表

M code	Function
M41	Low-gear range for spindle
M42	High-gear range for spindle
M48	Override cancel (OFF)
M49	Override cancel (ON)
M98	Transfer control to subroutine
M99	Return from a subroutine

表 B-2　用于车削操作的 M 代码(FANUC 控制器)汇总

M 指令	功能
M0	程序停止
M1	可选程序停止
M2	程序结束
M3	主轴顺时针旋转(CW)
M4	主轴逆时针旋转(CCW)
M5	主轴停止转动
M6	程序停止(自动换刀)
M8	冷却液开启
M9	冷却液关闭
M13	主轴顺时针/冷却液打开
M14	主轴逆时针/冷却液打开
M17	主轴关闭/冷却液关闭
M19	定向主轴停止
M30	程序结束,存储器复位
M41	主轴低档范围
M42	主轴的高速挡
M48	覆盖取消(关闭)
M49	覆盖取消(打开)
M98	子程序调用
M99	从子程序返回

Note：Many M codes are machine dependent and assigned by the machine tool builder. The machine tool manual should be checked for a complete list of M codes. For older controls, only one M code may be programmed in any block. If more than one is programmed, only the Fast-code is considered effective.

注意事项:许多 M 代码取决于机床并由机床制造商分配。应检查机床手册以获取 M 代码的完整列表。对于较旧的控制系统,在任何块中只能编程一个 M 代码。如果编程了多个,则只有快速代码才有效。

Appendix B Summary of G Codes, M Codes and Auxiliary Functions for Turning Operations
附录B　数控车的G代码、M代码和辅助变量汇总

Table B-3　Summary of Auxiliary Functions (FANUC Controllers)

code	Function
Dn	D is the address in memory where the cutter radius offset value is stored. O(xx) is the number of the register containing the offset value
Fn	F is the address in memory where the feed rate value is stored. n (xx.xxxx) is the numerical value of the feed rate
Hn	H is the address in memory where the tool length offset value is stored. n (xx) is the number of the register containing the offset value
In	Meaning varies depending upon G code. When used with G2/G3 circular interpolation (CW/CCW), I is the address in memory storing the incremental +(−) X distance from tool center to arc center with the tool positioned at the start of the arc cut. n (xx.xxxx) is the value of the distance. When used with G74 face grooving cycle, I is the address in memory storing the stepover/distance between grooves on the X-axis. n (xx.xxxx) is the value stepover/distance value
Jn	J is the address in memory storing the incremental +(−) Y distance from tool center to arc center with the tool positioned at the start of the arc cut. n (xx.Xxxx) is the value of the distance
Kn	Meaning varies depending upon G code. When used with G2/G3 circular interpolation (CW/CCW) in the XZ or YZ plane, I is the address in memory storing the incremental +(−)Z distance from tool center to arc center with the tool positioned at the start of the arc cut. n (xx.xxxx) is the value of the distance. When used with G74 face grooving cycle, K is the address in memory storing the first peck distance below the clearance plane on the Z-axis. n (xx.xxxx) is the value of the distance
Ln	L is the address in memory storing the number of times to repeat a call of a subprogram. n (xxxx) is the value of the number of times
Nn	N is the address in memory storing a sequence number in a program. n (xxxx) is the value of the sequence number
Pn	Meaning varies depending upon G code. When used with a G82 counterbore/spotface cycle, P is the address in memory storing the dwell time. n (xx.xxxx) is the value of the dwell time. When used with G71 stock removal or G70 finishing cycle, P is the address in memory storing the sequence number of the block that starts the contour description. n (xxxx) is the value of the sequence number
Qn	Meaning varies depending upon G code. When used with a G83 peck drill cycle, Q is the address in memory storing the first peck distance below the R_{plane}. n (xx.xxxx) is the value of the distance. When used with G71 stock removal or G70 finishing cycle, Q is the address in memory storing the sequence number of the block that ends the contour description. n (xxxx) is the value of the sequence number

续表

code	Function
Rn	Meaning varies depending upon G code. When used with aG2/G3 circular profile milling, R is the address in memory storing the radius of the tool path when the arc is cut. n (xx.xxxx) is the value of the radius. When used with G81–85 hole machining cycles, R is the address in memory storing the distance to the R_{plane}. n (xx.xxxx) is the value of the distance.
Sn	S is the address in memory storing the spindle speed. When used with G97, n (xxxx) is the value of the speed in revolutions per minute (r/min). When used with G96, n (xxxx) is the value of the speed in surface feet per minute (m/min).
Tab	T is the address in memory storing tool change values. a (xx) is the number of the turret station where the new tool is located. b (xx) is the address in memory where the new tool's offset values are stored
Xn	X is the address in memory where X-axis motion control is stored. n (xx.xxxx) is the value of the distance moved along the X-axis
Yn	Y is the address in memory where Y-axis motion control is stored. n (xx.xxxx) is the value of the distance moved along the Y-axis
Zn	Z is the address in memory where Z-axis motion control is stored. n (xx.xxxx) is the value of the distance moved along the Z-axis

表 B-3　辅助变量汇总（FANUC 控制器）

指令	功能
Dn	D 是存储刀具半径补偿值的存储器中的地址。 O(xx) 是包含补偿值的寄存器的编号
Fn	F 是存储器中存储进给量值的地址。 n(xx.xxxx) 是进给量的数值
Hn	H 是存储器中存储刀具长度补偿值的地址。 n(xx) 是包含补偿值的寄存器的编号
In	含义因 G 代码而异。 当与 G2/G3 圆弧插补（CW/CCW）一起使用时，I 是存储器中的地址，用于存储从圆弧中心到刀具起点的增量 +(-)X 距离。n(xx.xxxx) 是距离的值。 当与 G74 端面切槽循环一起使用时，I 是存储器中的地址，用于存储 X 轴上的凹槽之间的距离。n(xx.xxxx) 是指距离值
Jn	J 是存储器中的地址，存储从刀具起点到圆弧中心的增量 +(-)Y 距离。 n(xx.Xxxx) 是距离的值
Kn	含义因 G 代码而异。 当在 XZ 或 YZ 平面中使用 G2/G3 圆弧插补（CW/CCW）时，K 是存储器中的地址，用于存储从圆弧中心到刀具起点的增量 +(-)Z 距离。n(xx.xxxx) 是距离的值。 当与 G74 端面切槽循环一起使用时，K 是地址存储器在 Z 轴上切削深度。n(xx.xxxx) 是距离的值
Ln	L 是存储器中存储重复调用子程序的次数的地址。n(xxxx) 是次数的值

Appendix B Summary of G Codes, M Codes and Auxiliary Functions for Turning Operations
附录B 数控车的G代码、M代码和辅助变量汇总

续表

指令	功能
Nn	N 是存储器中存储序程序段号的存储器中的地址。n(xxxx)是程序段号的值
Pn	含义因 G 代码而异。 当与 G82 循环一起使用时，P 是存储停留时间的地址。n(xx.xxxx)是停留时间的值。 当与 G71 切削或 G70 精加工循环一起使用时，P 是存储器中的地址，用于存储开始轮廓描述的程序段的顺序号。n(xxxx)是序列号的值
Qn	含义因 G 代码而异。 当与 G83 啄式钻孔循环一起使用时，Q 是存储器中存储第一个切削深度的地址。n(xx.xxxx)是距离的值 当与 G71 切削或 G70 精加工循环一起使用时，Q 是存储器中的地址，用于存储精加工程序段的序列号，n(xxxx)是序列号的值
Rn	含义因 G 代码而异。 当与 G2/G3 圆形轮廓铣削一起使用时，R 是存储器中的地址，用于存储切削圆弧时刀具路径的半径。n(xx.xxxx)是半径的值。 当与 G81-85 孔加工循环一起使用时，R 是存储器中 R_{plane} 位置的地址。n(xx.xxxx)是距离的值
Sn	S 是存储主轴速度的存储器中的地址。 与 G97 一起使用时，n(xxxx)是以每分钟转数(r/min)为单位的速度值。 与 G96 一起使用时，n(xxxx)是以表面末每分(m/min)为单位的速度值
Tab	T 是存储刀具更改值的存储器中的地址。a(xx)是新工具所在的刀架的编号。 b(xx)是存储器中存储新工具偏移值的地址
Xn	X 是存储器中存储 X 轴运动控制的地址。n(xx.xxxx)是沿 X 轴移动的距离的值
Yn	Y 是存储器中存储 Y 轴运动控制的地址。n(xx.xxxx)是沿 Y 轴移动的距离的值
Zn	Z 是存储器中存储 Z 轴运动控制的地址 n(xx.xxxx)是沿 Z 轴移动的距离的值

Notes：

The format n (xx.xxxx) indicates

(1) The maximum range for the numerical value (the lowest range is one digit).

(2) That a decimal point (.) must always be coded with the number.

注意事项：

格式 n(xx.xxxx)表示

(1)数值的最大范围(最低范围是一位数)。

(2)小数点(.)必须始终用数字编码。

Appendix C Summary of Speeds and Feeds for Turning
附录C 车削加工过程中的车削速度和进给量汇总

Table C-1 Recommended Speeds for Turning

Turning Speeds（High-speed Steel Tools）

Material	Rough cuts	Average tool speed（m/min） Finish cuts
Magnesium	400	800
Aluminum	350	700
Brass and bronze	250	500
Copper	100	250
Cast iron（soft）	100	250
Cast iron（hard）	50	150
Mild steel	100	250
Cast steel	70	150
Alloy steel（hard）	50	150
Tool steel	50	150
Stainless steel	60	180
Titanium	90	200
High manganese steel	40	100

Note: For carbide cutting tools, double the average speed.

表 C-1 推荐的车削速度

车削速度(高速钢刀具)

材料	Rough cuts 粗车	平均刀具速度（m/min） Finish cuts 精车
镁	400	800
铝	350	700
黄铜和青铜	250	500
铜	100	250
铸铁（软）	100	250
铸铁（硬）	50	150
软钢	100	250
铸钢	70	150
合金钢(硬)	50	150
工具钢	50	150
不锈钢	60	180
钛	90	200
高锰钢	40	100

注意事项：对于硬质合金刀具，平均速度加倍。

Table C-2　Recommended Feeds for Turning

Turning Feeds

Material	Rough cuts	Finish cuts
		Tool feed (mm/r)
	Rough cuts	Finish cuts
Magnesium	0.38 ~ 0.63	0.12 ~ 0.25
Aluminum	0.38 ~ 0.63	0.12 ~ 0.25
Brass and bronze	0.38 ~ 0.63	0.07 ~ 0.25
Copper	0.25 ~ 0.50	0.10 ~ 0.20
Cast iron (soft)	0.38 ~ 0.63	0.12 ~ 0.25
Cast iron (hard)	0.25 ~ 0.50	0.07 ~ 0.25
Mild steel	0.25 ~ 0.50	0.07 ~ 0.25
Cast steel	0.25 ~ 0.50	0.07 ~ 0.25
Alloy steel (hard)	0.25 ~ 0.50	0.07 ~ 0.25
Tool steel	0.25 ~ 0.50	0.07 ~ 0.25
Stainless steel	0.25 ~ 0.50	0.07 ~ 0.25
Titanium	0.25 ~ 0.50	0.07 ~ 0.25
High manganese steel	0.25 ~ 0.50	0.07 ~ 0.25

表 C-2　推荐的车削进给

车削进给

材料	粗车	精车
		刀具进给 (mm/r)
镁	0.38 ~ 0.63	0.12 ~ 0.25
铝	0.38 ~ 0.63	0.12 ~ 0.25
黄铜和青铜	0.38 ~ 0.63	0.07 ~ 0.25
铜	0.25 ~ 0.50	0.10 ~ 0.20
铸铁（软）	0.38 ~ 0.63	0.12 ~ 0.25
铸铁（硬）	0.25 ~ 0.50	0.07 ~ 0.25
软钢	0.25 ~ 0.50	0.07 ~ 0.25
铸钢	0.25 ~ 0.50	0.07 ~ 0.25
合金钢（硬）	0.25 ~ 0.50	0.07 ~ 0.25
工具钢	0.25 ~ 0.50	0.07 ~ 0.25
不锈钢	0.25 ~ 0.50	0.07 ~ 0.25
钛	0.25 ~ 0.50	0.07 ~ 0.25
高锰钢	0.25 ~ 0.50	0.07 ~ 0.25

参 考 文 献

[1] 朱学超,刘旭,刘玉宏,等. 数控车床实训项目化教程[M]. 北京:机械工业出版社,2019.

[2] 张奇丽,李豪杰,胡建. 数控车削加工[M]. 重庆:重庆大学出版社,2015.

[3] 胡仲胜. 数控车床编程与操作[M]. 重庆:重庆大学出版社,2015.

[4] 郁汉琪. 数字化设计与制造实训教程[M]. 南京:东南大学出版社,2016.

[5] 贺泽虎. 数控车床编程与加工应用实例[M]. 重庆:重庆大学出版社,2015.

[6] 章继涛,田科,刘井才,等. 数控技能训练[M]. 北京:人民邮电出版社,2014.

[7] 周湛学,刘玉忠. 数控编程速查手册[M]. 北京:化学工业出版社,2013.

[8] 包德莹. 数控机床编程与操作[M]. 沈阳:辽宁科学技术出版社,2009.

[9] 杜军. 数控宏程序编程手册[M]. 北京:化学工业出版社,2014.

[10] EMADI A, KHALIGH A, NIE Z, et al. Integrated Power Electronic Converters and Digital Control[M]. CRC Press, 2017.

[11] VELONI A, MIRDAKIS N. Digital Control Systems:Theoretical Problems and Simulation Tools[M]. CRC Press, 2017.